Clinical ENT

An Illustrated Textbook

· ·

GERARD M. O'DONOGHUE
Consultant ENT Surgeon, University Hospital, Nottingham

GRANT J. BATES
Consultant ENT Surgeon, The Radcliffe Infirmary, Oxford

AND

ANTONY A. NARULA
Consultant ENT Surgeon, The Leicester Royal Infirmary, Leicester

with a Foreword by
SIR DONALD HARRISON

Oxford   New York   Tokyo
OXFORD UNIVERSITY PRESS

Oxford University Press, Walton Street, Oxford OX2 6DP

Oxford New York
Athens Auckland Bangkok Bombay
Calcutta Cape Town Dar es Salaam Delhi
Florence Hong Kong Istanbul Karachi
Kuala Lumpur Madras Madrid Melbourne
Mexico City Nairobi Paris Singapore
Taipei Tokyo Toronto

and associated companies in
Berlin Ibadan

Oxford is a trade mark of Oxford University Press

Published in the United States by
Oxford University Press Inc., New York

A catalogue record for this book is available from the British Library

Library of Congress Cataloging in Publication Data
O'Donoghue, Gerard M.
Clinical ENT : an illustrated textbook / Gerard M. O'Donoghue,
Antony A. Narula, and Grant K. Bates.
Includes bibliographical references.
1. Otolaryngology. I. Narula, Antony A. II. Bates, Grant J. III. Title.
[DNLM: 1. Otorhinolaryngologic Diseases-atlases. WV 17 026c]
RF46.O33 1992 91–31839
ISBN 0-19-261667-6 (pbk.) ISBN 0-19-262226-9

Typeset by
Footnote Graphics
Warminster, Wilts
Printed and bound in Hong Kong

Foreword

by Professor Sir Donald Harrison

One of the most rewarding features of my three decades as an academic otorhinolaryngologist has been the meteoric rise of this specialty's status and scope. Despite recognition that over one third of all referrals to general practitioners relate to ENT disorders, most medical students receive scanty indoctrination into the mysteries of the specialty. This leaves them uninformed and unprepared when faced with 'real' medical practice, and they will surely welcome this excellent book with considerable relief. Hopefully, medical students will also become aware of its value, thus avoiding a painful learning experience.

Many smaller, more superficial attempts have been made to fill this educational void, but to my knowledge this is the first occasion that a triumvirate of enthusiastic, well-trained young teachers have pooled their experience to produce a reasonably sized book specifically designed to cover comprehensively all aspects of the modern specialty. To do this in 29 well-illustrated chapters is in itself an achievement, but to inform without being tedious is a considerable accomplishment. It is much harder to write a book where much interesting detail must be excluded, and the manner in which this has been attained in this publication amply reflects the thought and care exercised by these young men. They are a credit to this thriving specialty and many will thank them for this masterly contribution to otorhinolaryngology and head and neck surgery.

Professor Sir Donald Harrison by Mr Kelvin E. Thomas F.R.C.S.

Preface

The challenge the authors faced was this: how to convey the excitement of this bustling specialty to the contemporary medical student? The development that has characterized ENT in recent years is extremely impressive. Much has to be attributed to advances in sensory physiology with which our specialty is intimately linked. For instance, developments in auditory neurophysiology have revolutionized the assessment of inner ear disorders and paved the way for such important advances as cochlear implantation. The skull base, previously a surgical 'no man's land', is now eminently accessible using microsurgical techniques. Endoscopic surgery has had a major impact on the understanding and treatment of nasal and sinus disorders and has advanced the evaluation and surgery of the larynx and trachea. Better reconstructive techniques have become available to minimize the morbidity from major head and neck resections. The use of lasers and microvascular techniques has been especially valuable. Voice disorders, cosmetic surgery, sleep disturbance, and molecular biology are examples of other areas engaging much of the ENT surgeon's attention.

Care has been taken to cover the basics of ENT assessment and common clinical conditions are presented in a succinct manner, often illustrated by clinical photographs. Realizing that medical students too have their pressures, key points have been interspersed throughout the text to emphasize the essentials and help with the inevitable last minute revision!

We also hope the book will help family practitioners, casualty officers, paediatricians, and the many physicians whose practice brings them to bear on that great temple of surprise—the head and neck.

Nottingham   G.M.O'D.
Oxford   G.J.B.
Leicester   A.A.N.
November 1991

Acknowledgements

The authors are indebted to their many colleagues and students who helped in the preparation of this book, especially those who courageously offered their criticism!

Many illustrations and photographs have been generously loaned to us by our friends. The eminent French otologist, Dr Christian Deguine from Lille readily put his remarkable collection of tympanic membrane photographs at our disposal. Mr Michael Rothera F.R.C.S., Manchester, kindly provided Figs 115, 116, 117, 123, and 124, Dr David Kennedy M.D., Baltimore (Fig. 145), Mr Andrew Freeland F.R.C.S., Oxford and Oxford University Press (Fig. 5), Mr Michael Stearns F.R.C.S., London (Figs 241 and 242), Mr J. Oates F.R.C.S., Nottingham (Fig. 270); Dr Simon Allison F.R.C.P., Nottingham (Fig. 224); Dr Steve Mason, Nottingham (Figs 33, 34, 75, 105), and Shiley (UK) Ltd (Figs 313 and 316). We are indebted to the Medical Illustration Departments at University Hospital, Nottingham and the Leicester Royal Infirmary for their contributions. Thanks go to Susan Bates for all the illustrations in Part 2. We would like to thank the staff of Oxford University Press for their unfailing help, understanding, and great patience.

A special word of thanks to Raphaële, Susan, Charlotte, and the children—without their understanding, patience, and sense of humour, we might have faltered in our mission.

Contents

PART 3 · THE LARYNX, HEAD, AND NECK

The Ear

CHAPTER ONE

Clinical anatomy and physiology

The ear is a sensory end organ serving the requirements of hearing and balance.

THE EXTERNAL EAR

The external ear (Fig. 1) helps collect and localize sound. It is formed by the pinna and external ear canal. The pinna is made of elastic cartilage covered by skin. The skin is tightly adherent to the perichondrium on the outer surface of the pinna. A haematoma may detach the perichondrium and devascularize the cartilage. Deficiencies in the cartilage of the ear canal can facilitate the spread of infection and malignancy to the parotid and skull base.

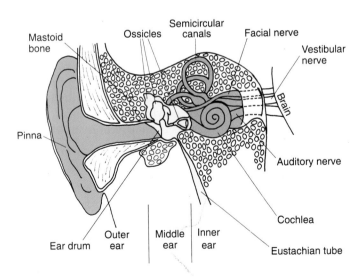

Fig. 1 A coronal section through the ear. Sound is collected by the external ear, amplified by the middle ear and converted into electrical impulses by the inner ear. Note that the facial nerve courses through the ear.

In the adult, the external ear canal is between 2 and 3 cm in length. The outer two-thirds is cartilaginous and the inner third is bony. The ear canal is slightly curved and to allow inspection must be straightened by gently retracting the pinna upwards and backwards (Fig. 2).

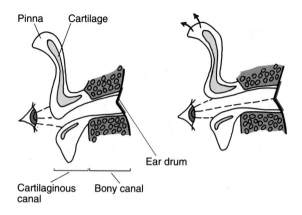

Fig. 2 To inspect the tympanic membrane the ear canal must be straightened. This is done by gently retracting the pinna upwards and backwards.

The inner two-thirds of the ear canal is lined by highly specialized stratified squamous epithelium, which is devoid of hair follicles and does not desquamate. This skin has the property, unique in the body, of lateral migration. This is an active process, starting at the ear drum, and progresses at the rate of about 100 μm per day. This 'conveyor belt' keeps the ear canal clean, which is essential for efficient air conduction. Attempts by patients to clean their ears with cotton buds only serve to push wax and debris further down the ear canal and such practices should be discouraged. The skin of the outer third of the ear canal is identical to skin elsewhere on the body.

Wax is formed by ceruminous glands in the outer portion of the ear canal. It prevents the ingress of particular matter into the ear and also has a surface immunoprotective function. It only needs to be removed if it becomes impacted (usually in people who clean their ears). The oily property of this wax was used by ancient monks to illuminate their manuscripts!

The *lymphatics* of the external ear drain to the retroauricular, parotid, retropharyngeal and upper deep cervical lymph modes. These may become enlarged and tender in infections or neoplasms of the external ear.

THE MIDDLE EAR

The *tympanic membrane* (Figs 3 and 4) or ear drum separates the external and middle ears. It has a dense fibrous middle layer and an inner layer lined by middle ear mucosa. Its small upper portion, above the lateral process, is called the *pars flaccida*, because its middle fibrous layer is relatively lax. The lower portion is called the *pars tensa* and is the part of the drum responsive to sound. In health it is virtually transparent and exhibits a *light reflex* on otoscopic examination (this is due to

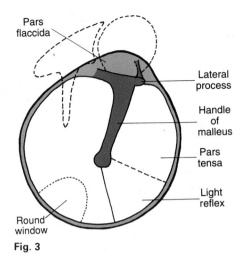

Fig. 3

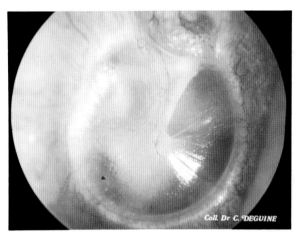

Coll. Dr C. DEGUINE

Fig. 4

its slightly conical shape). The malleus handle can be seen embedded in the ear drum. The tympanic membrane measures about 1 cm in diameter and its centre is called the umbo.

The middle ear is an air-containing space bounded laterally by the tympanic membrane and medially by the basal turn of the cochlea, called the promontory. Its capacity is about 1 cm^3. It contains a chain of three bones or *ossicles*: the malleus which is the largest and is embedded in the tympanic membrane, the incus in the middle and the stapes lying in contact with the inner ear at the oval window. The incus is delicately poised and is the ossicle most likely to be dislocated following head injury. Tiny muscles attach to the malleus and stapes to dampen vibration and protect the inner ear from excessive noise.

The middle ear is important because infection can spread through it to vital structures only a few millimetres away. Superiorly lies the middle cranial fossa (containing the temporal lobe). Posteriorly, the mastoid air cells are adjacent to the posterior fossa (containing the cerebellum) and the lateral (sigmoid) sinus. Medially, the lateral semicircular canal can be eroded by chronic suppuration which allows bacteria access to the inner ear. The *facial nerve* runs through the middle ear and can also be damaged by inflammatory processes.

The Eustachian tube connects the middle ear to the nasopharynx. Oxygen is constantly absorbed by the mucosa of the middle ear, resulting in a negative middle ear pressure. The Eustachian tube is responsible for the equalization of pressure between the middle ear and the outside world. The tube is normally closed but has muscles attached to it which open the tube on swallowing or yawning. Its function is particularly important during flying and diving to enable rapid equalization of pressure on both sides of the tympanic membrane. The tube undergoes major structural changes during development: in infancy it is short and

Fig. 3 Normal tympanic membrane (*right*). The conical shape of the membrane causes the light reflex. Note the position of the lateral process and handle of malleus. The pars flaccida lies above the lateral process.

Fig. 4 Normal membrane (*left*). Compare this with Fig. 3. Note its semi-transparent quality. The silhouette of the long process of incus and round window niche can be appreciated through the intact membrane.

almost horizontal but during childhood becomes elongated, with a slight downward angulation. The relative immaturity of the tube contributes to the high prevalence of childhood middle ear disease.

The inner ear structures are immersed in fluid. The transmission of sound energy from air (i.e. as in the middle ear) to a fluid medium results in substantial loss of sound energy (Fig. 5). To overcome this, the middle ear needs to amplify sound. This is achieved by virtue of the fact that the surface area of the tympanic membrane is 20 times that of the oval window. Furthermore, the shape of the chain of ossicles is such as to confer a mechanical advantage for sound transmission.

THE INNER EAR

The inner ear comprises the cochlea, vestibular labyrinth, and their central connections. These delicate structures are embedded in dense bone called the otic capsule. The facial nerve courses through the inner ear.

Fig. 5 The goldfish is unperturbed by the alarm clock as most of the airborne sound is reflected at the air-water interface. Without the amplifying action of the middle ear in man, almost all of the incident sound energy would be reflected away from the inner ear.

The cochlea

The cochlea is a minute spiral of two and a half turns. Within this spiral, perilymph and endolymph are partitioned by the thinnest of membranes. Biochemically, the inner ear fluids are quite different, with endolymph having a high concentration of potassium (i.e. 144 mEq/l, similar to intracellular fluid) and perilymph being high in sodium (Fig. 6). The maintenance of these ionic concentrations is an active process effected by sodium and potassium pumps.

To understand how this complex structure works, it is best to unravel its spiral arrangement (Fig. 7). The endolymphatic compartment containing the sensory epithelium is completely surrounded by perilymph. Each compression caused by the stapes footplate results in a corresponding rarefaction of the round window membrane. A travelling

Fig. 6 The biochemical concentrations of inner ear fluids. Cerebrospinal fluid is similar to perilymph with which it communicates. Endolymph has a high potassium concentration similar to intracellular fluid.

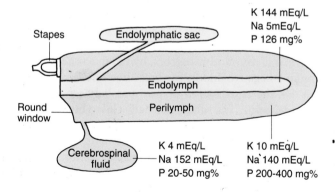

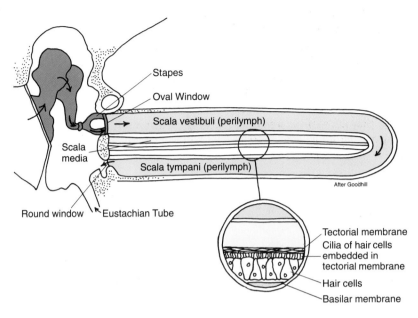

Fig. 7 The spiral of the cochlea is undone. Each compression at the stapes causes a rarefaction at the round window (the inner ear fluid being incompressible). A travelling wave is generated which maximally stimulates certain parts of the basilar membrane: high frequencies at the base of the cochlea, low frequencies at the apex.

wave is generated which causes maximum resonance in the particular portion of the basilar membrane which roughly corresponds to the particular stimulus frequency. The lower frequencies register in the apical turn, the higher frequencies in the basal turn. Although minute, the cochlea accommodates an intensity range of 1 000 000:1. For reasons of practicality, this vast range is compressed into a logarithmic decibel scale of 0 to 120 dB (Fig. 8). The frequency range of the human ear is from 20 to 20 000 Hz.

The basic physiological unit in the cochlea is the *hair cell* (Fig. 9). There are about 15 000 such cells in the human ear. They are arranged in rows of inner and outer cells. The hair cells act as mechano-electric transducers converting the acoustic signal into an electrical impulse (Fig. 10). The slightest displacement of the hairs causes rapid ionic shifts to occur, with depolarization of the cell followed by an immediate return to the resting potential. This electrical impulse is carried from hair cell to neuron by release of a chemical neurotransmitter. Such is the sensitivity of hair cells that they respond within a few millionths of a second to movements of 100 picometers, which is no more than the diameter of an atom!

The auditory nerve carries impulses to the cochlear nuclei in the brain stem. Each nerve fibre is 'tuned' to respond best to a particular

Fig. 8 The decibel scale. A whisper is about 20 dB, conversational voice about 60 dB, and a jet aircraft 120 dB. The scale is logarithmic to accommodate the vast intensity range of the human ear.

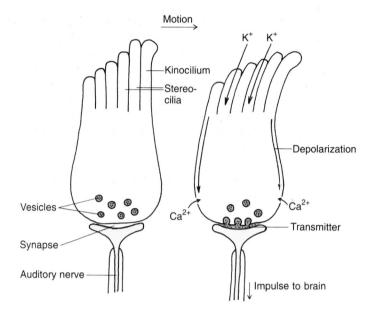

Fig. 9 A hair cell. The slightest motion (e.g. the diameter of a hydrogen atom!) of the hairs causes rapid depolarization of the cell. This results in the release of an unknown neurotransmitter which excites the auditory neuron. The impulse is then carried along the auditory nerve.

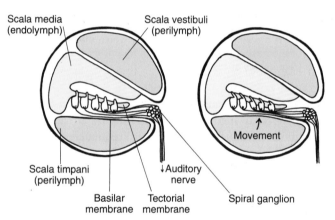

Fig. 10 Movement of the basilar membrane in response to sound results in a shearing of the hair cells which are embedded in the tectorial membrane. Impulses generated in this way are then carried via the auditory nerve.

frequency, giving it a characteristic 'tuning' curve. Most fibres cross to the opposite side of the brain stem and travel to the lateral lemniscus and medial geniculate body. From there, the fibres travel to the auditory cortex which is located in the superior temporal gyrus of the cerebral cortex.

The vestibular labyrinth

This comprises the semicircular canals, the utricle, saccule, and their central connections. The three semicircular canals (horizontal, superior, and posterior) are arranged in the three planes of space at right angles to each other.

Like the auditory system, the vestibular end organ responds to hair cell movement (Fig. 11). In the lateral canals, the hair cells are embedded in a gelatinous cupula and are sheared during angular movements of the head. In the utricle and saccule, the hair cells are embedded in an otoconial membrane containing particles of calcium carbonate. These respond to changes in linear acceleration and the pull of gravity. Impulses are carried centrally by the vestibular nerve. The major connections of the vestibular system are to the spinal cord, cerebellum, and external ocular muscles.

The maintenance of balance is dependent not only on the integrity of the vestibular apparatus but also on input from proprioceptive and visual systems and the cerebellum.

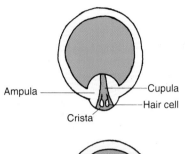

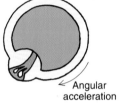

Key Point

The hair cells of the inner ear act by converting sound and acceleration into electrical energy which excites the auditory and vestibular nerves.

SENSORY INNERVATION OF THE EAR

External Ear
 Great auricular (C2 and C3), Lesser occipital (C2)
 Auriculo temporal (Cranial nerve V)
 Sensory branch of Facial (Cranial nerve VII)
 Vagus (Cranial nerve X)

Middle Ear
 Glossopharyngeal (Cranial nerve IX)

Inner Ear
 No somatic sensory innervation

Referred otalgia may come from the normal area of distribution of any of these nerves

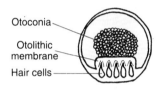

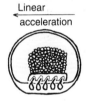

Fig. 11 Angular and linear acceleration cause shearing of hair cells in the semicircular canals, utricle, and saccule. Impulses generated in this way are carried by the vestibular nerve to the cerebellum, spinal cord, and the nuclei of the ocular muscles.

Key Point

Referred pain to the ear may come from many structures in the head and neck (e.g. a sore throat or impacted wisdom tooth). Always examine the oral cavity and pharynx to exclude infection or tumour.

CHAPTER TWO

History and examination

....

The ear is a superficial structure on the side of the head and can be readily inspected. It is disappointing to find that examination is so often inadequate or not undertaken at all. One should not be deterred by the mystique of the head mirror or microscope!

HISTORY

Table 1. The five principal symptoms of ear disease

1. Earache
2. Deafness
3. Discharge
4. Tinnitus
5. Vertigo

Diseases of the ear can only express themselves in a very limited number of ways (Table 1).

Earache (otalgia)

The ear has a rich and varied sensory innervation (p. 9). Earache usually indicates an inflammatory process in the external or middle ear. The onset of the pain, its duration, severity, aggravating and relieving factors, as well as the presence of any associated symptoms need to be determined. The quantity of analgesics taken by the patient and the amount of time lost from school or work are useful pointers to the severity of the pain. Referred pain may come from many head and neck structures (wisdom teeth, jaw joint, floor of mouth, tonsil, cervical spine, etc.)

Deafness

There are four key questions.

Was the onset sudden or gradual? Sudden hearing loss may occur in a number of circumstances, for example following barotrauma (flying or diving), upper respiratory infection, exposure to excessive noise, drug administration (e.g. gentamicin), or head injury.

Is it unilateral or bilateral? Remember that unilateral losses are more likely to have a specific underlying cause. Deafness due to ageing or noise exposure is not unilateral. The commonest presenting symptom of an acoustic neuroma is unilateral hearing loss which is often very slight and insidious. Bilateral hearing loss may have a genetic basis or be due to systemic disease.

Is the loss progressive, fluctuating, or improving? Deafness due to ageing or continued noise exposure is often progressive. Meniere's disease and serous otitis media typically present with fluctuating hearing loss. An acute middle ear effusion following an upper respiratory infection usually shows steady improvement.

What is the hearing like in group conversation? Generally, patients with conductive hearing loss manage reasonably well in group conversation. Patients with sensorineural hearing loss have poor speech discrimination in noise and usually admit to doing better in 'one to one' conversation. This question will also help assess the impact of the hearing impairment on the patient's ability to communicate in everyday life.

Discharge (otorrhoea)

- Diseases of the external ear produce a scanty discharge
- Copious discharge is always from the middle ear
- Smelly discharge suggests cholesteatoma
- Bloody discharge follows ear canal trauma, severe acute infection or tumour
- Watery discharge can be cerebrospinal fluid

> **Key Point**
> Mucoid discharge must come from the middle ear as there are no mucous glands in the external ear.

Tinnitus

Tinnitus is the name given to the symptom of noises in the head or ears. Enquiry should be made about the circumstances of onset, the type of noise, its location, duration, and intensity. Tinnitus that is pulsatile may be due to vascular abnormality or tumour. Often the noises are most troublesome at night when the patient is trying to get to sleep. This is due to the absence of background noise. Tinnitus is a common accompaniment of hearing impairment from any cause.

Vertigo

This term describes the hallucination of movement of the environment about the patient or of the patient in relation to the environment. It is *not* synonymous with dizziness. Vertigo is a symptom of a disorder of the peripheral labyrinth or of the central vestibular pathways. The issues to determine in relation to vertigo are as follows:

Onset. If the symptom started following barotrauma (flying or diving), acoustic trauma or head injury, rupture of an inner ear membrane (perilymph fistula) should be considered. Acute fulminant vertigo without hearing loss in an otherwise healthy young adult is likely to be due to vestibular neuronitis. If the symptom lasts a few seconds on adopting a sudden change of posture without any hearing loss, consider benign positional vertigo (BPV). Dysequilibrium (rather than vertigo) lasting over many years may have a psychogenic basis.

Duration. In Meniere's disease, the vertigo lasts between half an hour and ten hours. In BPV, the vertigo lasts for only a few seconds. In vestibular neuronitis, the duration is two to three days.

Associated symptoms. The presence of auditory symptoms strongly supports a peripheral labyrinthine cause and should always be asked about. Nausea and vomiting, as in sea sickness, are typical accompaniments of labyrinthine vertigo. Central causes of vertigo (e.g. vertebrobasilar insufficiency, multiple sclerosis) are associated with other neurological symptoms (e.g. dysarthria, visual disturbance) and are not typically accompanied by auditory symptoms. Limitation of neck movement may indicate cervical vertigo.

EXAMINATION

Instruments

The instruments required for examination of the ear are shown in Fig. 12.

Technique

Position. The examination is best performed when the examiner and the patient are seated comfortably (Fig. 13). It is distressing to see a patient seated while the physician stands, screwing his spinal column into contortions and perilously grasping an otoscope (Fig. 14). This is likely to be both unproductive and hazardous. Therefore, sit comfortably at the patient's level.

Inspection. Inspect the pinna. Having done so, bend it forwards and look behind the ear (Fig. 15). This may reveal the scarring of previous surgery, a swelling which is often inflammatory or the ulceration typical of malignancy. Bat ears is the term given to abnormally protuberant ears. Deformity of the pinna may be a clue to congenital hearing loss.

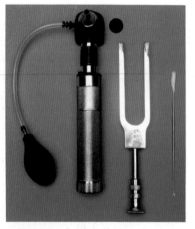

Fig. 12 Tools of the trade! From left to right: a fibreoptic otoscope with disposable speculae (note the pneumatic attachment of the otoscope with which the mobility of the ear drum can be tested); a tuning fork (512 Hz); a Jobson Horne probe (one end can act as a cotton wool carrier and the other is used for wax removal).

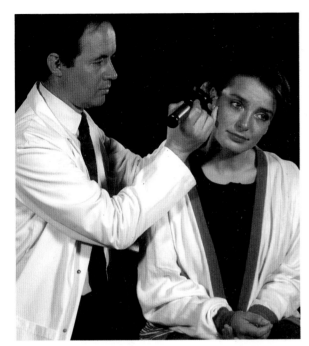

Fig. 13 When patient and examiner are at the same level, the examination is greatly facilitated.

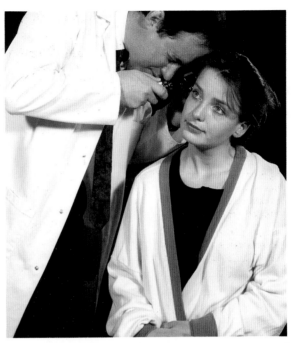

Fig. 14 This awkward posture has nothing to recommend it. Note the poor control over the otoscope which is being held like a dagger.

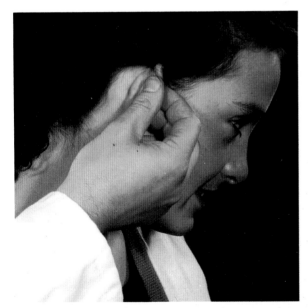

Fig. 15 This manoeuvre should be done in every case. It is surprising what the pinna can hide—scars, swellings due to infection, or tumours.

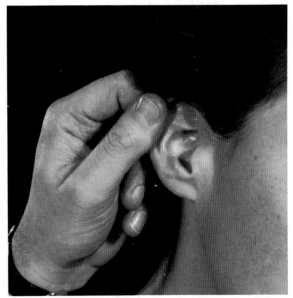

Fig. 16 Retracting the pinna upwards and backwards straightens the external ear canal and is a prerequisite for inspecting the tympanic membrane.

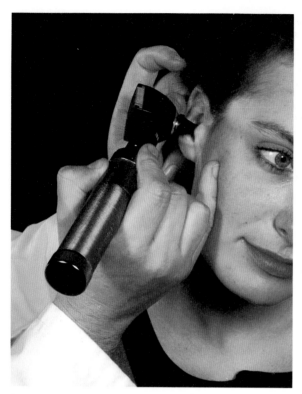

Fig. 17 Always have control over any instrument (otoscope, wax hook, cotton wool carrier, etc.) entering the external ear. The little finger resting on the cheek is the best 'early warning' system. This is the arrangement for the right ear.

Fig. 18 Examination of the left ear needs a change of hands.

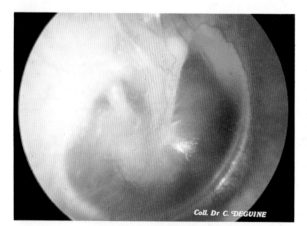

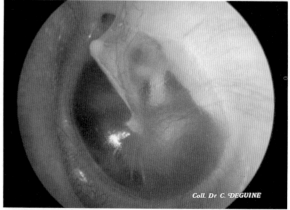

Fig. 19 Normal tympanic membrane (*right*). Note its semi-transparent quality. The silhouette of the long process of incus and round window niche can be appreciated through the intact membrane.

Fig. 20 Normal tympanic membrane (*left*). Compare the landmarks on the right (Fig. 19) and left ear drums. The lateral process of the malleus always points anteriorly.

Otoscopy. Hold the otoscope like an artist holds a brush—delicately but with full control. *Gently* retract the pinna to straighten the ear canal (Fig. 16)—bear in mind that in inflammatory conditions of the ear this movement alone can be painful. Watch the patient's facial expression—if there are signs of grimacing, proceed only with the greatest caution. Always steady the hand which holds the instrument on the patient's cheek (Figs 17 and 18). This is essential to prevent injury to the external ear should the patient suddenly jolt. If wax is present, it will need removal by syringing.

Once the ear canal has been cleared, it is essential to examine the entire tympanic membrane (Figs 19 and 20). There is a tendency to satisfy oneself with seeing the pars tensa but the pars flaccida is more often the seat of destructive ear disease. To examine the pars flaccida, direct the light at the most superior aspect of the ear drum while tilting the patient's head away.

The mobility of the ear drum can be determined by using the pneumatic attachment on the otoscope (Fig. 12), provided the speculum is a snug fit for the ear canal (which is not often the case with contemporary equipment!).

Tuning forks. Tuning fork testing must be included in the assessment. The tests demand good technique and a basic understanding of air and bone conduction. The tests most commonly employed are the *Rinne and Weber tests.*

The *Rinne test* compares hearing through air and bone conduction. Normally, we hear by air conduction which is more efficient than hearing through bone. To do the test, briefly explain to the patient that you are going to place a tuning fork behind and then in front of the ear and that you will want to know in which position the fork sounds loudest. Strike the tuning fork (512 Hz) against your elbow or knee cap (not on the floor or table top—this produces undesirable overtones). Then steady the patient's head with one hand and apply the base of the tuning fork firmly to the patient's mastoid bone for at least two to three seconds, i.e. sufficient time for the patient to make a mental note of the intensity of the stimulus (Fig. 21). Then bring the tuning fork around to the external ear canal (Fig. 22) and again allow a few seconds for the patient to make a judgement. In a normal patient or in a patient with sensorineural hearing loss, the Rinne test is positive, i.e. loudest in front of the ear. The Rinne test is said to be negative when the fork sounds loudest behind the ear. This is typical of conductive hearing loss greater than 20 decibels.

A *false negative Rinne test* can be a pitfall for the unwary. If a patient has no cochlear function on one side, the Rinne test will be negative because the good ear (and *not* the test ear) is picking up the sound by means of bone conduction. To avoid this happening, the non-test ear must be masked ('kept busy') during testing, i.e. by massaging the tragus. The Weber test will usually save the day by being referred to the only hearing ear.

In the *Weber test* the tuning fork is placed in the midline (usually at the vertex or on the bridge of the nose) and the patient is asked to indicate in which ear, if any, the noise sounds loudest. Again, explain the procedure to the patient before you proceed with the test. Normally, the sound is perceived in the mid-line. If one ear has a sensorineural deficit, the sound lateralizes to the better cochlea. If a conductive loss is present on one side, the sound goes to the side of the conductive loss (Figs 23 and 24).

Conversational testing. Conversational testing offers a useful guide to the level of hearing impairment. It must be done without giving the patient any visual clues such as the examiner's lip movements. In many situations, the non-test ear will need to be prevented from hearing. This is done by simply massaging the tragus of the non-test ear while simultaneously asking the patient to repeat words or numbers spoken near the ear under test. To a rough approximation, a whisper is about 30 decibels; a softly spoken voice is about 50 decibels.

Nystagmus. Nystagmus is involuntary oscillatory movement of the eyes. It can be a normal phenomenon. If you sit opposite someone who is looking out of the window in a fast moving train you will observe the eyes flicker every few seconds. This nystagmus has two components: a quick component due to central compensation and the slow phase which is due to peripheral mechanisms. By convention, the direction of a nystagmus is the direction of the fast component. Peripheral nystagmus tends to be suppressed by optic fixation and hence is enhanced in darkness or by wearing Frenzel's glasses (Fig. 25). These glasses have +20 dioptre lenses which allow the examiner a magnified view of the eye movements but prevent the patient from using optic fixation.

Horizontal jerk nystagmus is typical of peripheral labyrinthine disorders. This has a slow movement in one direction followed by a quick jerk in the opposite direction. Nystagmus due to central disorders, i.e. multiple sclerosis or brain stem compression, tends to be much more complex and often bizarre.

Nystagmus induced by positional testing is a useful test in patients with benign positional vertigo. The patient is asked to sit on an examining couch and the head is thrust suddenly backwards to just below the level of the couch (Figs 26 and 27). In benign positional vertigo nystagmus is induced when the affected ear is lowermost. The nystagmus comes on after a second or so and lasts for several seconds. It is of a rotatory type and is fatiguable, i.e. it disappears with repeated testing. The test is not a way to win friends but is clinically very useful!

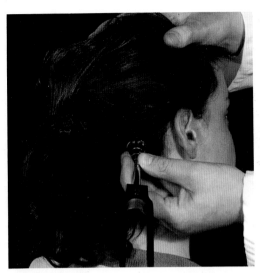

Fig. 21 The Rinne test—bone conduction. To establish good bone contact, first steady the patient's head with one hand and then firmly apply the tuning fork to the mastoid bone.

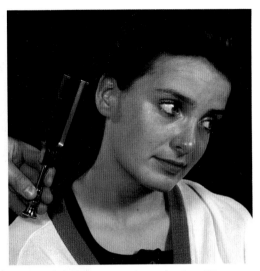

Fig. 22 The Rinne test—air conduction. When bringing the tuning fork to the 'front', ensure the tines are parallel to the external ear canal.

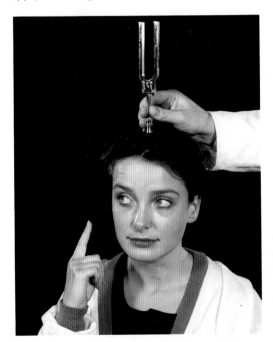

Fig. 23 The Weber test. Place the tuning fork firmly on a bony point in the mid line (vertex, bridge of nose, incisor teeth). Weber to the right: the patient may rotate the eyes to the right or point to the right ear. This happens typically with a right conductive hearing loss.

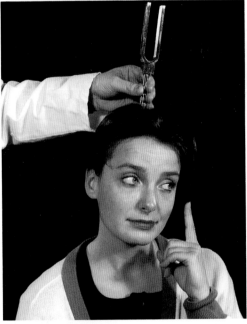

Fig. 24 The Weber test lateralizing to the left ear (e.g. left conductive hearing loss). The Weber test is much more sensitive than the Rinne test.

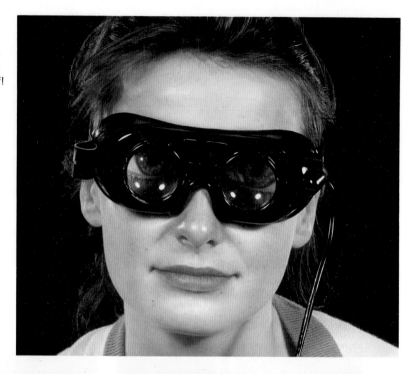

Fig. 25 Frenzel's glasses. These remove optic fixation by having +20 dioptre lenses—try them on yourself! They magnify the eyeball and make eye movements much easier to define. Removing optic fixation makes nystagmus from inner ear disease much more obvious.

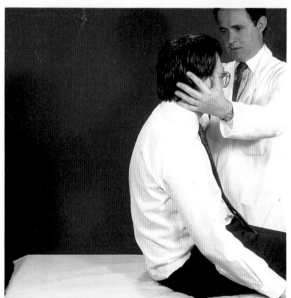

Fig. 26 Positional testing. The starting position must be such as to allow the patient's head to be dropped below the horizontal when the patient is thrust backwards (Fig. 27).

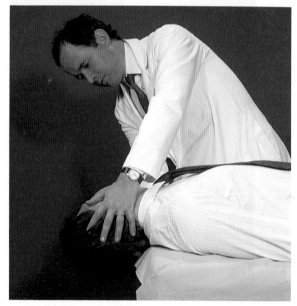

Fig. 27 Positional testing. Note that the head is below the level of the couch and rotated to one side. The patient's eyes should be open to observe nystagmus. The test should not be done in patients with restricted neck mobility.

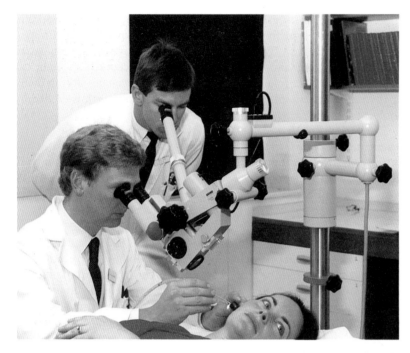

Fig. 28 Examination under the microscope (EUM). This allows inspection under magnification of the external ear, middle ear, or mastoid cavity. Ears containing debris or wax can be suction cleaned under a microscope. A myringotomy (drainage of fluid from the middle ear) can be performed in clinic using this technique under local anaesthetic.

Table 2. ENT: history-taking and examination

Ear	Nose	Throat
History		
Earache, irritation	Obstruction	Hoarseness
Deafness	Rhinorrhoea/post nasal drip	Dysphagia
Discharge	Allergy/hay fever	Stridor
Tinnitus	Facial pain	Lump in the neck
Vertigo	Epistaxis	
	Sense of smell	Sleep disturbance
	Appearance	
Children: Speech/language		
Past History		
Barotrauma	Trauma	Cigarette smoking
Acoustic trauma	Medications	Alcohol
Head injury	*prescribed*	
Ototoxics	*non prescribed*	
Family history	Previous surgery	
Previous ear surgery		
Examination		
Pinna	Shape	Mouth
Mastoid	Septum	Larynx—refer to ENT
Ear canal	Turbinates	Neck
Tympanic membrane	Airway	
Tuning forks	?Mucopus	
Conversational test	Facial tenderness	
Nystagmus	Facial sensation	
Facial nerve	Facial swelling	

Key Points

1. Conductive hearing losses are due to disorders of the external or middle ear.

2. Sensorineural hearing losses are due to disorders of the cochlea or of the auditory nerve.

3. Tuning forks tests can determine whether a hearing loss is conductive or sensorineural and should always be performed. They will not indicate the level of hearing loss.

4. The Rinne test will not become negative until a conductive hearing loss is in excess of 20 decibels.

Fistula test. This is performed by raising and lowering the pressure in the external ear (usually with a pneumatic otoscope or by a pumping action on the tragus). The test is said to be positive if the manoeuvre induces nystagmus and vertigo. This typically occurs if an inner ear membrane ruptures following excessive pressure changes during flying or diving (barotrauma). It also occurs when the labyrinth is eroded by cholesteatoma.

Facial nerve. The facial nerve is motor to the facial muscles and to the stapedius muscle in the middle ear. Always check facial nerve function in diseases of the ear. Usually the diagnosis is obvious but a subtle weakness (see Figs 95 and 98 on p. 55) or a bilateral paralysis (see Figs 93 and 94 on p. 55) can easily escape detection.

Examination under the microscope (EUM). This is standard procedure in all ENT departments (Fig. 28). A speculum is placed in the external ear canal and the pinna is retracted. Using this technique, ears can be dewaxed, foreign bodies removed, and minor surgery (e.g. myringotomy or grommet insertion) can be undertaken. A valuable adjunct is the use of suction to remove debris from an ear.

The main steps in the examination of the ear are summarized in Table 2.

CHAPTER THREE

Investigation of auditory, vestibular, and facial nerve function

PURE TONE AUDIOMETRY

Pure tone audiometry is the test of auditory function most frequently undertaken. It is best performed in a soundproof booth (Fig. 29). The test is dependent on patient co-operation and it is important that time is spent explaining to the patient what is involved. The patient wears head phones through which pure tones are presented to each ear in turn. The patient is asked to respond to sounds of decreasing intensity across the frequency range. Both air conduction and bone conduction can be tested (Fig. 30). Bone conduction is a measure of cochlear function.

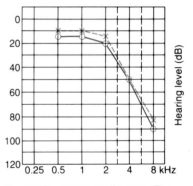

Fig. 30 Pure tone audiogram. The zero decibel line represents the normal threshold. Hearing deficits are recorded below this line. The low frequencies are on the left with high frequencies on the right. Agreed symbols are used for right and left ears and for air and bone conduction.

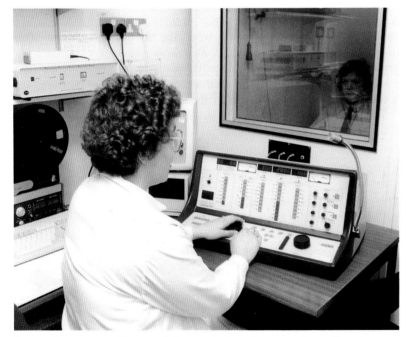

Fig. 29 Pure tone audiometry being carried out in a sound proof booth. Sounds of decreasing intensity are presented and the patient is asked to press a trigger each time he hears a sound. No visual or temporal cues must be given to the patient about the sounds.

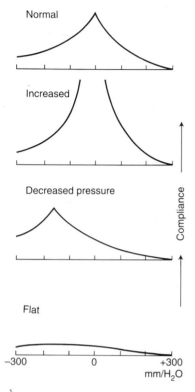

Normal

Increased

Decreased pressure

Flat

Compliance

−300   0   +300
mm/H$_2$O

Fig. 31 The four most common types of tympanogram: *normal, increased compliance, decreased compliance, and flat.*

Fig. 32 A normal electrocochleogram (above top). Note the changes typical of an excess of fluid in the endolymph compartment, as occurs in Meniere's disease (above bottom).

Acoustic separation of the ears can be a very difficult task. When testing visual acuity, the tester can simply occlude the eye which is not being tested—this cannot be done so readily for hearing. The non-test ear must be masked ('kept busy') by applying a noise of appropriate intensity.

SPEECH AUDIOMETRY

In this test the patient is presented with phonetically balanced words and is scored on the number of correct responses.

TYMPANOMETRY

Tympanometry is a technique which is used extensively in clinical practice. It is based on the fact that the pressure in the external ear can be raised or lowered, thus stiffening the ear drum. A tympanometer presents a low frequency sound to the ear and measures the sound energy reflected from the ear drum. The ear drum is most 'floppy' when the pressure on both sides of it is equal. If fluid is present in the middle ear, the ear drum is unresponsive to changes in pressure of the external ear canal and a 'flat' tracing is observed.

The four most common types of tympanogram are shown in Fig. 31 *Normal*: occurs when pressures on both sides of the drum are equal at 0 on the pressure scale. *Increased*: when the ear drum is 'floppy' (thin scars, ossicular chain disrupted). *Decreased pressure* occurs when the pressure in the middle ear is negative relative to the external ear (e.g. Eustachian tube dysfunction). *Flat*: occurs in middle ear effusions and is due to the fact that the fluid in the middle ear is incompressible.

Tympanometry is most useful in screening children for otitis media with effusion. The equipment is affordable in general practice.

STAPEDIAL REFLEXES

The stapedius muscle contracts in response to sound. This contraction stiffens the ossicular chain. This change in stiffness is readily detectable by measuring the resistance of the ossicular chain to sound transmission. In otosclerosis, a condition where the stapes bone is fixed, the stapedial reflex is absent. Scarring in the middle ear, caused by otitis media, can also result in loss of the reflex.

ELECTRIC RESPONSE AUDIOMETRY

This is the most important development in modern audiometry. Remember the anatomy of the central auditory pathways. Impulses originate in the cochlea and are carried by auditory nerve fibres to a number

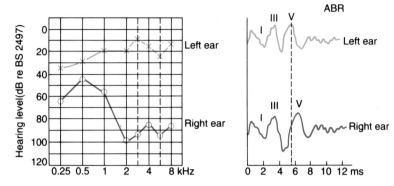

Fig. 33 Pure tone audiogram and auditory brainstem responses in a patient with a right sided acoustic neuroma. The hearing is normal in the left ear as is the I–V latency (broken vertical line represents upper limit of normal). On the right side, there is high frequency hearing loss and the I–V latency is prolonged. (*Note* Wave I—distal cochlear nerve, Wave III—cochlear nuclei, and Wave V—tracts and nuclei of the lateral lemniscus).

of relay stations in the brain stem before reaching the auditory cortex. By means of surface recording electrodes on the head and computer averaging techniques it is possible to study the response of the auditory pathways to sound.

If a single discrete sound is presented to the auditory system, the electrical response to it will be 'drowned' by background activity from the brain (EEG) and muscles. If, however, thousands of time-locked discrete sounds are presented (rather like machine-gun fire), the electrical responses they generate can be added up to give a recognizable tracing. The background electrical activity in the brain, being random, will then cancel out. No anaesthesia is generally required—the patient can even nod off to sleep during the study! There are three different types of electric responses in common usage:

1. *Electrocochleography (ECochG)*. This measures physiological events in the cochlea in response to sound (Fig. 32). In Meniere's disease, specific changes are seen because the excess of fluid interferes with the mechanical properties of the membranes in the cochlea.

2. *Auditory brainstem responses (ABR)*. This resembles a nerve conduction study. The time taken for an impulse to get from the cochlea and through to the brain stem is measured. This time is prolonged by acoustic neuromas, even when they are tiny. Hence the key role of this investigation in screening patients for these tumours (Fig. 33). The test can be used as an objective hearing test in babies and young children. None of these two tests provides frequency specific information as in each case a broad bandwidth stimulus is used.

3. *Auditory Cortical Responses*. This technique measures the electrical activity in the auditory cortex in response to sounds of different

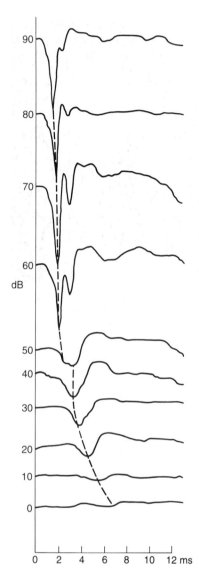

90

80

70

60

dB

50

40

30

20

10

0

0 2 4 6 8 10 12 ms

Fig. 34 Auditory cortical responses in a normal ear. Note the waveforms can be traced from 90 dB almost to 0 dB.

frequencies (Fig. 34). The responses are particularly useful when assessing hearing in those patients who are either unable or unwilling to perform conventional audiometry. Thus, it is particularly useful in patients who may be exaggerating a hearing deficit in the pursuit of compensation.

OTOACOUSTIC EMISSIONS

Within the last decade it has been discovered that hair cells in the inner ear actually make sounds, called emissions, which can be detected by placing a highly sensitive microphone in the ear canal. They are a reliable indicator of the integrity of the peripheral auditory system. The emissions are absent in patients with mild inner ear deafness. They are being evaluated in the screening of neonates for sensorineural hearing loss.

CALORIC TESTING

The function of the semicircular canals is assessed by observing the effect on the system of warm and cold water stimulation. The vestibular nuclei have direct connections with the oculomotor nuclei. In the test, cold (30 °C) and warm (44 °C) water are introduced alternately into the external ear. This sets up convection currents in the lateral semicircular canals which stimulate the vestibular system, producing nystagmus (involuntary eye movements). On cold water stimulation, the direction of the nystagmus (i.e. that of the fast component) is to the opposite side while warm water induces a nystagmus to the same side (the mnemonic COWS—Cold Opposite, Warm Same is helpful). The duration of the nystagmus is measured and the responses from left and right sides are compared.

ELECTRONYSTAGMOGRAPHY

This technique allows the caloric response to be recorded graphically. It uses the fact that the eye is a rotating di-pole, the retina being negative relative to the cornea. Recording electrodes are placed adjacent to the orbit and can record eye movements even when the eyes are closed (i.e. with removal of optic fixation).

ELECTRONEURONOGRAPHY (ENoG)

This is used in the assessment of patients with facial paralysis. The facial nerve is electrically stimulated by placing electrodes on the skin above its exit point from the temporal bone (i.e. just below the ear lobe).

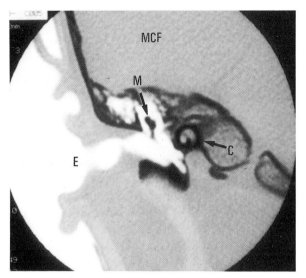

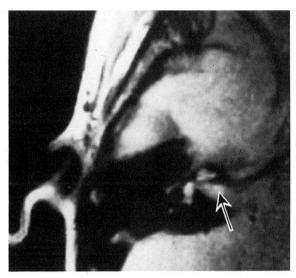

Fig. 35 CT scan right ear. This is a coronal section. Note the external ear canal (E), tympanic membrane and malleus (M), a coil of the cochlea (C), and the middle cranial fossa (MCF).

Fig. 36 MRI. This is a horizontal section. The auditory nerve is well visualized and contains a tiny acoustic neuroma (arrowed). Note the poor visualization of the surrounding bone.

Recording electrodes are placed over the facial muscles. If Wallerian degeneration has taken place, no response will be detected. This has important implications for prognosis and in deciding if a nerve ought to be surgically explored.

RADIOLOGY

1. *Plain films*. Plain films and tomography of the temporal bone have almost been rendered obsolete by CT scanning.
2. *Computed tomographic (CT) scanning*. High resolution sections 1 to 2 mm thick enable visualization of the detailed bony anatomy of all parts of the ear (Fig. 35). Soft tissues, such as cholesteatoma or acoustic neuroma, can also be clearly seen on these views.
3. *Magnetic resonance imaging (MRI)*. This complementary method of investigation has overcome many of the limitations of CT scanning. It is better at imaging soft tissue (such as the auditory or facial nerve) but is poor at defining bone detail. It is excellent at imaging tiny tumours arising from the auditory nerve (acoustic neuromas) (Fig. 36) or affections of the brain stem. Its images can be enhanced by using contrast agents such as gadolinium.

CHAPTER FOUR

Common problems of the external ear

The external ear is a skin-lined structure adapted for sound collection. It comprises the pinna and external ear canal.

AURICULAR HAEMATOMA

This typically occurs following blunt injury to the side of the head as may happen in a rugby scrum. The skin of the pinna is tightly bound to the perichondrium of the underlying elastic cartilage. The cartilage depends on perichondrium for its nutrition. A haematoma may detach the perichondrium (Fig. 37) and cause necrosis of the underlying cartilage, especially if the haematoma becomes infected (perichondritis).

Treatment

Early incision and drainage with strict aseptic technique, usually under anaesthesia, and followed by a pressure dressing for one week. Broad spectrum antibiotic cover is recommended.

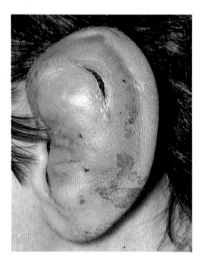

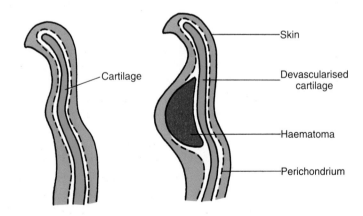

Fig. 37 Auricular haematoma. Such a haematoma can devascularize the cartilage.

CAULIFLOWER EAR

If the treatment of perichondritis (Fig. 38) is neglected (early intravenous antibiotics are essential), the cartilage may shrivel giving rise to this unsightly deformity.

CONGENITAL MALFORMATIONS

These are dealt with in the section on paediatric otology (Chapter 8).

IMPACTED WAX

This is a frequent occurrence in general practice, especialy in elderly patients. The use of cotton buds to clean the ear usually causes accumulation of wax deep in the meatus, often impacting it against the ear drum. The hearing loss is usually slight, except when the meatus is totally obstructed.

When the wax is hard, soften it by using warm olive oil drops or sodium bicarbonate ear drops for two to three weeks. Wax removal is rarely an emergency and it is often kinder and safer to do it in stages. Proprietary preparations (cerumolytics) can also be used but they sometimes irritate the skin of the external ear. If wax remains after this therapy, the ear can usually be syringed successfully.

Syringing

Injury to the ear is frequent with this technique, usually because insufficient care is taken in its execution (Figs 39–42). Seat the patient or have him lie on a couch and describe the procedure. *Ask if there is any history of perforation of the ear drum.*

Check the temperature of the water with a thermometer—it should be warmed precisely to body temperature to avoid caloric stimulation of the labyrinth. Check the syringe—the tip should be blunt and securely attached to the barrel of the syringe. As few patients appreciate being drenched, a protective towel around the neck is desirable.

Steady the patient's head. Gently retract the pinna upwards and backwards. Syringing must always be done under direct vision. Position the syringe adjacent to the ear canal and inject the water with firm, even pressure. *Direct the flow of water to the back wall of the meatus.* This protects the ear drum from the full force of the stream of water. It also allows the water to get behind the mass of wax and will enable it to be dislodged and flushed outwards. After the procedure, dry mop the ear canal with a wisp of cotton wool.

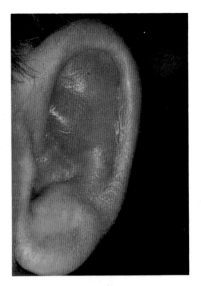

Fig. 38 Perichondritis. The ear felt hot and was painful. Note the diffuse erythema and swelling. Systemic antibiotics prevented cartilage destruction.

Key Point

Syringing is contraindicated following recent injury or in patients with a history of perforation of the ear drum.

Fig. 39 Instruments for ear syringing. A jug for warm water, syringe, thermometer, cotton tip applicator, different sized tips for the syringe, and a receptacle to collect the water.

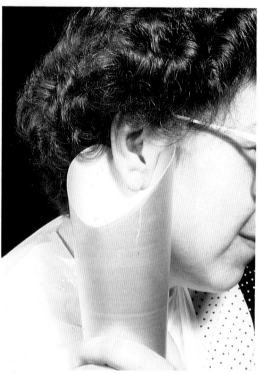

Fig. 40 A specially designed receptacle to collect water during ear syringing in place underneath the ear.

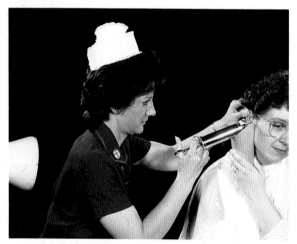

Fig. 41 Syringing technique. A head lamp is ideal but, as shown here, an anglepoise lamp will suffice. The syringe is directed *towards the back wall* of the ear canal to protect the ear drum from the full force of the jet of water.

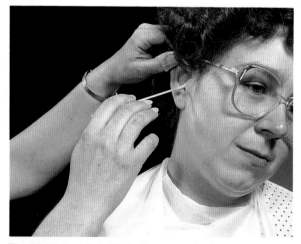

Fig. 42 Water left behind after ear syringing can be very uncomfortable. Dry mopping the ear canal is recommended.

OTITIS EXTERNA

This is one of the commonest affections of the ear. It can be very troublesome to treat if some basic rules are not followed. Otitis externa can be reactive (i.e. due to contact allergy) or infective (bacterial, viral, fungal).

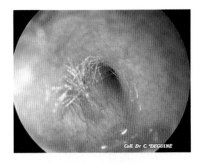

Fig. 43 An oedematous ear canal in otitis externa. A microscope is usually required to suction clean such an ear and to insert a dressing with safety.

Symptoms

The cardinal symptom is *irritation*. It can be intense and often affects both ears simultaneously. Ask about a history of eczema or contact allergy to chemicals, shampoos, or cosmetics. Certain occupations, such as telephonists who require inserts in their ears, are often affected. Pain occurs only when secondary bacterial infection sets in and hearing loss, if present, is usually mild. Discharge is scanty (there are no mucous secreting glands in the external ear).

Treatment

The first consideration is *aural toilet*. Until this is satisfactory, all treatments are doomed to failure. Gentle dry mopping with a wisp of wool on a wool carrier is best—be extremely gentle as the external ear is exquisitely tender. Syringing is best avoided in an acutely inflamed canal. If the ear canal cannot be cleaned or is very narrowed by swelling (Fig. 43), specialist help may be needed to clean the ear with suction using a microscope.

Once clean, conventional wisdom dictates that a combination of *steroids and antibiotic drops* be given to settle the ear for a week or so. Bear in mind that some patients may be allergic to these drops (Fig. 44). Prolonged use of antibiotics can give rise to superinfection with yeasts (Figs 45 and 46) and should be avoided.

One old-fashioned remedy, insertion of a glycerine and ichthammol wick (Fig. 47), has much to recommend it. The glycerine is hygroscopic (draws water from the oedematous canal), the ichthammol is soothing and mildly antiseptic. The wick (3 to 4 inches of ½ inch ribbon gauze) helps get the solution deep into the canal. Commercially available otowicks (Fig. 48) that swell when moistened with ear drops, are also very helpful with narrowed ear canals.

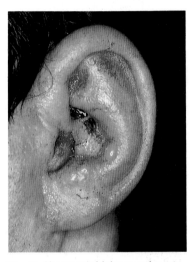

Fig. 44 Hypersensitivity reaction to ear drops. The pinna is oedematous and there is blister formation.

Advise the patient to keep the ear canal dry as water aggravates otitis externa. Cotton wool smeared with vaseline is an excellent means of achieving this.

Finally, the patient must be persuaded to stop scratching the ear canal with matchsticks, cotton buds or finger nails (Fig. 49). The rule is: *nothing smaller than the elbow in the external ear!*

Some hints on the management of refractory cases of otitis externa are outlined in Table 3.

Key Points

1. Profuse discharge implies middle ear disease—not otitis externa.

2. Persistent unilateral 'otitis externa' equals otitis media until proven otherwise.

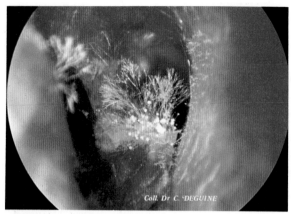

Fig. 45 Fungal otitis externa. Note the fluffy appearance given by the hyphae. This may follow prolonged treatment with antibacterial ear drops.

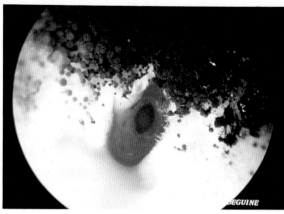

Fig. 46 The striking appearance of clumps of Aspergillus niger in the ear canal. Once seen, never 'forgotten!

Fig. 47 Wicks much loved by ENT surgeons. On the left is BIPP (bismuth iodoform paraffin paste—contraindicated in patients with iodine allergy), and on the right is glycerine and ichthammol (GI). A useful angled aural dressing forceps is shown.

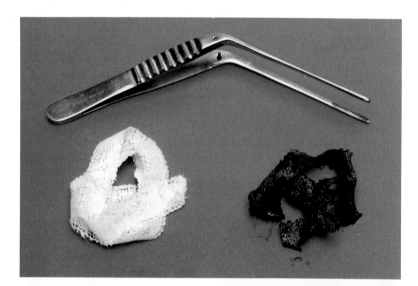

Fig. 48 An otowick. Prior to insertion, it is firm and narrow (left). When drops are applied (left) it softens and expands (right). These are easy to insert even in narrow ear canals (Fig. 43).

Fig. 49 Instruments of torture for the ear canal! Finger nails, cotton buds, hair clips, and matchsticks are frequently used to scratch the ear canal and cause secondary infection.

Fig. 48

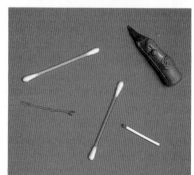

Fig. 49

Table 3. The stubborn otitis externa

Problem	Comment
Aural toilet	If the ear canal is not clean, do not waste time and money changing to yet another brand of ear drops. Either clean the ear or refer to ENT for microsuction.
Fungal infection	Follows prolonged treatment with antibacterial ear drops. Clean the ear and treat with nystatin powder or clotrimazole drops. Fungal otitis externa is often indistinguishable from bacterial otitis externa except for *Aspergillus niger* (Fig. 46) whose hyphae form striking black clusters on the canal wall.
Sensitivity to ear drops	Remember that a patient may develop an allergy to ear drops (Fig. 44). This can be dramatic and very painful.
Hearing aids	The mould of a hearing aid can alter the humidity of the external ear and predispose to otitis externa. The use of a hearing aid in the acutely inflamed ear is best discouraged. Fit the hearing aid in the other ear where possible and allow the inflammation to subside. A hole can be made in the ear mould to allow ventilation of the ear canal in troublesome cases. Allergy to the ear mould material should be investigated.
Middle ear disease	Otitis externa can be secondary to middle ear disease. Suspect in unilateral cases and those with profuse discharge.
Microbiology	Should be obtained in difficult cases. Ask for culture for fungi and tuberculosis.

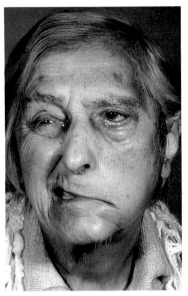

Fig. 50 Necrotizing 'malignant' otitis externa. This elderly diabetic developed otitis externa and facial paralysis. Despite intensive medical and surgical treatment, she succumbed to the disease.

NECROTIZING ('MALIGNANT') OTITIS EXTERNA

This destructive inflammatory process affects immune compromised patients, especially elderly diabetics (Fig. 50). The term 'malignant' is unfortunate but was applied to this condition to denote its high mortality. It starts as a stubborn otitis externa which gradually causes an osteomyelitis of the skull base. Pain is often severe. Several cranial nerves may be paralysed, especially the facial nerve and nerves IX to XII. The offending organisms are *Pseudomonas aeruginosa* and anaerobes.

Investigations

Swabs for culture (remember the anaerobes); CT, MRI, or technetium scans.

Treatment

Intensive local therapy as well as systemic antibiotics against *Pseudomonas* (ciprofloxacin or ceftazidime being especially effective) and anaerobes.

Key Point

Beware of otitis externa in diabetic patients.

FURUNCLE OF THE EXTERNAL EAR

This is an infection of a hair follicle, rather like a boil, in the outer third of the ear canal (Fig. 51). It is excruciatingly painful.

Treatment

Be extremely gentle. Local treatment to the swollen ear canal (glycerine and ichthammol wick or otowick (Figs 47 and 48)) along with intensive systemic antibiotic therapy against staphylococci.

FOREIGN BODIES

Foreign bodies often find their way into the ear canal (Fig. 52) and can be very difficult to remove. What may at first sight seem a trivial technical challenge can soon end up as an inelegant fiasco. The danger is of pushing the object (beads are the worst!) further into the ear canal causing damage to the ear drum and middle ear. This can easily happen with a fractious child. Vegetable matter can swell and may impact during syringing.

The rule is that patients with objects that are not readily amenable to removal need referral to an ENT department.

EAR CANAL TRAUMA

This is usually caused by the insertion of cotton buds and hair clips in the ear (see also skull base trauma p.49). The result is a laceration of the canal and bleeding. It may not be possible to see the ear drum to exclude injury to the middle or inner ear. The treatment is usually one of masterly inactivity (Figs 53 and 54). The blood clot is best left undisturbed as it acts as a wound dressing. Further evaluation should be carried out after a few weeks.

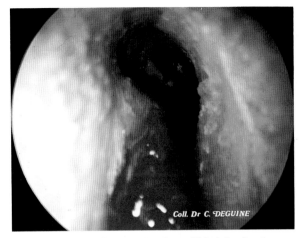

Fig. 51 Furuncle of the external ear. This infection in a hair follicle of the external ear is excruciatingly painful and the canal is exquisitely tender. In this patient, an abscess has ruptured into the meatus.

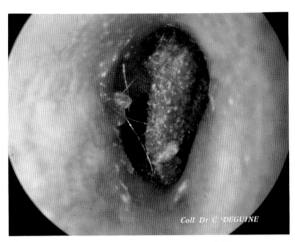

Fig. 52 Foreign body deep in the ear canal. These are best removed by an ENT surgeon.

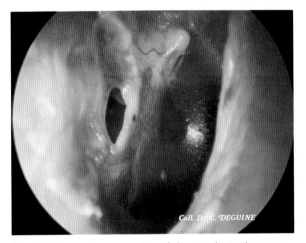

Fig. 53 Traumatic perforation of the ear drum due to a cotton tip.

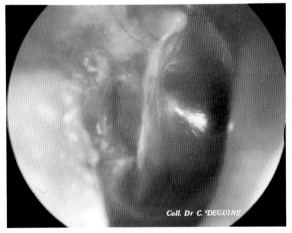

Fig. 54 The same ear as in Fig. 53. With masterly inactivity the perforation healed spontaneously in a matter of weeks.

TUMOURS

Pinna

The commonest tumour is a *squamous cell carcinoma* (Fig. 55) which has an ulcerated appearance with raised everted edges. Excessive exposure to sunlight is a predisposing factor. This gives rise to solar keratoses which are pre-malignant. These lesions require early ENT assessment

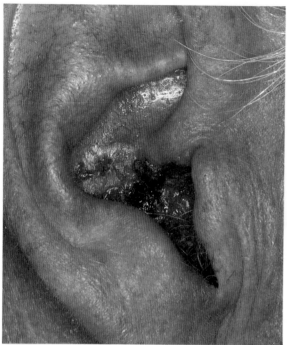

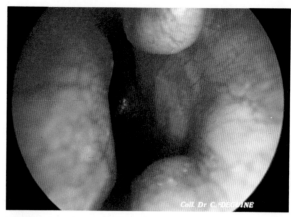

Fig. 56

Fig. 55 This squamous carcinoma of the pinna had raised everted edges. A more common site is on the outer surface of the pinna, especially in those exposed to sunlight.

Fig. 56 Osteomas of the ear canal. These occur in swimmers. They do not require treatment unless they block the ear canal.

Fig. 55

Fig. 57 Squamous cell carcinoma of the ear canal. The patient was treated for many months for 'otitis externa'. The pain was intense and facial paralysis set in. The patient's facies conveys her distress.

as the prognosis is excellent when the disease is localized. Metastases usually go to the neck. *Basal cell carcinomas* (rodent ulcers) may also affect the pinna. They have a 'punched out' appearance. Local invasion rather than distant metastasis is the rule.

Ear canal

These are uncommon. Benign tumours typically arise from the ceruminous glands. Osteomas (Fig. 56) may arise from frequent swimming in cold water and need no treatment unless they obstruct the canal.

Malignant primary growths are rare and are usually squamous cell carcinomas. Sometimes the ear canal may be invaded by tumours in adjacent structures, usually the parotid. Deep pain, sometimes accompanied by bloody discharge or facial paralysis, is the hallmark of malignancy (Fig. 57). Resectable malignant tumours are treated primarily by surgery with or without the addition of radiation therapy.

Common conditions of the middle ear

The middle ear contains the chain of bones necessary for sound conduction (Fig. 58). Laterally, it is bounded by the tympanic membrane, medially by the inner ear and anteriorly by the Eustachian tube. The roof is a wafer thin bone which separates the middle ear from the middle fossa dura. Behind, the middle ear communicates with the mastoid air cell system, which is closely related to the posterior cranial fossa. Entwined in this tiny space is the facial nerve, which pursues a tortuous course through the middle ear.

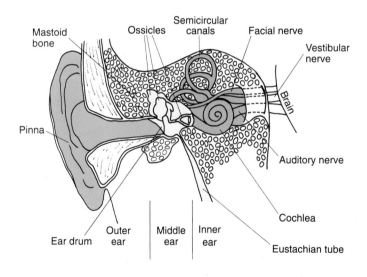

Fig. 58 The middle ear and its immediate relationships. Note the thin layer of bone (the tegmen tympani) which separates the middle ear from the middle cranial fossa.

The following conditions are described in the chapter on paediatric otology:

Acute otitis media (see p. 60)

Recurrent otitis media (see p. 60)

Acute mastoiditis and masked mastoiditis (see p. 62)

Otitis media with effusion (chronic serous otitis media, glue ear) (see p. 63)

MIDDLE EAR EFFUSION IN ADULTS

Typically these follow upper respiratory tract infection or barotrauma. Patients with nasal obstruction (polyps, allergic, or vasomotor rhinitis) may also develop fluid in their middle ears. Unilateral effusion in an adult should always be viewed with suspicion. A nasopharyngeal tumour may invade the muscles surrounding the Eustachian tube and thus prevent its opening. This applies especially to southern Chinese patients in whom such tumours are endemic.

CHRONIC SUPPURATIVE OTITIS MEDIA

There are two varieties: 'safe' and 'unsafe'. The latter is more important as it refers to a destructive process. It is usually due to cholesteatoma. The 'safe' variety implies a simple chronic discharge from the ear without the destructive sequelae typical of cholesteatoma.

CHRONIC SUPPURATIVE OTITIS MEDIA—'UNSAFE'

Cholesteatoma

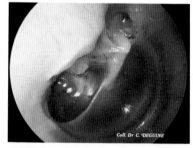

Fig. 59 Retraction pockets. These form when the tympanic membrane gets 'sucked in' to the middle ear. Gradually the pocket deepens and fills with desquamating epithelium. Compare with Fig. 60.

This is the term given to keratinizing squamous epithelium in the middle ear. The cause is unknown. It is possible that persistent negative pressure in the middle ear causes the tympanic membrane to be sucked inwards resulting in retraction pocket formation (Figs 59 and 60). The skin desquamates causing expansion of these pockets which invariably become infected. These 'skin bags' are locally destructive to all the structures in the temporal bone. As for *symptoms*, smelly discharge from an ear is the hallmark of this condition. Invariably, the discharge is accompanied by hearing loss. If the inner ear is eroded, the patient may become vertiginous or lose hearing completely. Pressure on the facial nerve can cause paralysis of the face.

Spread of infection outside the ear may result in intracranial suppuration (see p. 39), usually affecting the temporal lobe.

Examination

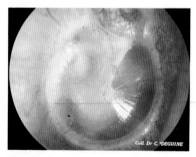

Fig. 60 A normal tympanic membrane for comparison.

Otoscopy reveals debris in the ear canal. Look with care at the attic area (the uppermost part of the ear drum) in these cases. Here you will see either a crust (Fig. 61) or a perforation full of whitish debris (Fig. 62). Tuning forks usually indicate a conductive hearing loss. Culture from the discharge usually grows *Pseudomas aeruginosa* and anaerobes.

Investigations

Examination under the microscope, audiometry; microbiology and CT scans are occasionally indicated.

Treatment

Once the diagnosis is made (and it is easy to do so) it is pointless to persist with ear drops and antibiotics. Surgical treatment, usually a mastoid exploration, is necessary to treat the condition.

The main priority for the surgeon is to remove the diseased area and render the patient safe from the threat of continued suppuration. Following mastoid surgery the ear should be dry and trouble free. Regular follow up will be needed for life as cholesteatomas can recur. To improve hearing, reconstruction of the ossicular chain (ossiculoplasty) may be feasible at a later stage. Provided the ear is dry following surgery, the patient also has the option of wearing a hearing aid. A patient with a persistently discharging ear following previous surgery should be offered a revision procedure.

> **Key Points**
>
> **1.** Smelly discharge from the ear strongly suggests cholesteatoma.
>
> **2.** Suppurative ear disease is still the commonest cause of temporal lobe abscess.
>
> **3.** A discharging mastoid cavity is no safer than a discharging ear. These patients should be considered for revision surgery.
>
> **4.** The main risks of mastoid surgery are sensorineural hearing loss, vertigo, and facial paralysis.

CHRONIC SUPPURATIVE OTITIS MEDIA— 'SAFE'

In this condition, the problem is caused by perforation of the pars tensa (Fig. 63). External and nasopharyngeal infection can easily gain access to the middle ear, irritating the mucosa and causing increased mucus production—it is rather like chronic bronchitis in the ear! Unlike cholesteatoma, the discharge is not offensive but is just as copious. Typically, there is hearing impairment.

Examination

Otoscopy reveals a perforation in the pars tensa (Figs 64 and 65). Tuning fork testing will suggest a conductive hearing impairment.

Investigations

Audiogram, microsuction, culture, and sensitivity (occasionally).

Treatment

Medical management (debridement, antibiotic/steroid drops, keeping the ear dry) can often resolve matters to the patient's satisfaction. A hearing aid, which works well for conductive losses, may need to be prescribed. If medical therapy fails or if the patient is keen on surgery (for instance, to allow swimming or for occupational reasons) then the defect can be grafted (myringoplasty). With any operation on the ear there is a small risk of sensorineural hearing loss and the enthusiasm for surgery (either by surgeon or patient) should be tempered by this consideration.

TYMPANOSCLEROSIS

This term is applied to 'chalk patches' in the tympanic membrane or middle ear (Fig. 66). A horse-shoe shaped ring of tympanosclerosis often follows grommet insertion. The process starts as hyalin degeneration followed by calcification. When tympanosclerosis is confined to the tympanic membrane it is rarely associated with hearing loss.

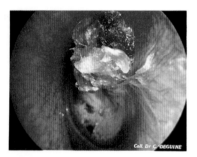

Fig. 61 An attic crust. Beneath such a crust usually lurks a cholesteatoma.

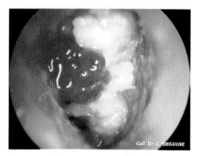

Fig. 62 Cholesteatoma. It was originally described as a 'pearly tumour' due to its whitish, glistening appearance. Note the granulation tissue which often accompanies cholesteatoma.

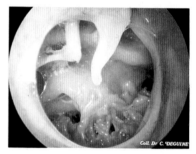

Fig. 63 A large pars tensa perforation (right ear). The malleus handle, long process of incus, stapedius muscle, round window niche, and promontory can be seen.

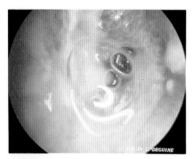

Fig. 64 Chronic suppurative otitis media—'safe'. During an exacerbation mucoid discharge fills the ear canal.

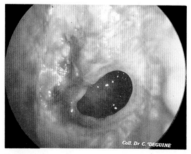

Fig. 65 Chronic suppurative otitis media—'safe'. The same ear as in Fig. 64 following microsuction. A central perforation through which can be seen an angry-looking, hyperaemic middle ear mucosa.

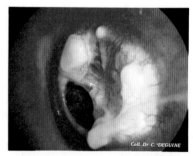

Fig. 66 Tympanosclerosis ('chalk patches'). Degenerative changes can result in plaques of calcification in the tympanic membrane or middle ear.

TUBERCULOUS OTITIS MEDIA

An important cause of suppuration in developing countries. The diagnosis should be considered in those ears failing to respond to standard therapy, especially in patients from third world countries with multiple perforations of the ear drum. A swab for appropriate culture studies coupled with a chest X-ray will usually confirm the diagnosis.

INTRACRANIAL COMPLICATIONS OF EAR INFECTION

The most serious complication of acute or chronic suppuration in an ear is intracranial sepsis (Figs 67 and 68). The most important warning signs are headache and deep pain in the ear (Table 4). Uncomplicated chronic ear disease is not painful. Patients with such a history should be sent for urgent ENT assessment.

Management

Patients should be managed in consultation with a neurosurgeon. A CT scan of the brain is essential. Lumbar puncture should not be performed in patients with raised intracranial pressure. A brain abscess will need repeated aspiration and intensive antibiotic therapy. The definitive treatment of the underlying ear disease is best carried out when the patient's condition has stabilized.

Key Points

1. Earache or headache in a patient with chronic ear disease should suggest the possibility of intracranial complication.

2. A normal CT scan does not exclude intracranial suppuration.

3. Early neurosurgical consultation should be obtained for patients with suspected or established intracranial disease.

4. Suspect disease in the ear (or nose) in a patient with recurrent attacks of bacterial meningitis.

OTOSCLEROSIS

This condition affects the dense otic capsule bone which houses the inner ear. New spongy bone is formed in the region of the stapes footplate (Fig. 69). This inhibits the mobility of the footplate resulting in conductive hearing loss. The condition is usually bilateral and symmetrical. It is inherited as a Mendelian autosomal dominant trait with

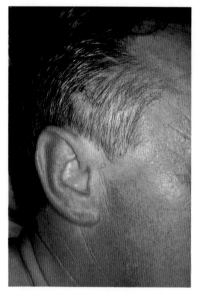

Fig. 67 Intracranial complication of ear disease. This patient presented to casualty with headache and a swelling above the ear which was incised. Within a few hours of leaving hospital, he was unconscious (see Fig. 68).

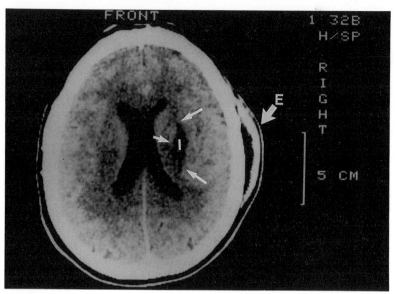

Fig. 68 Intracranial complication of ear disease. The same patient as in Fig. 67. CT scan. There is an abscess in the temporal lobe (I) as well as a large collection of pus extracranially (E). The source of infection was cholesteatoma.

Table 4. Intracranial suppuration from ear disease

Diagnosis	Clinical features
Meningitis	Neck stiffness, photophobia, positive Kernig's sign, etc.)
Lateral sinus thrombosis	Headache, rigors, spiking temperature, papilloedema, positive blood culture
Temporal lobe abscess	Headaches, drowsiness, visual field defects, dysphasia etc.
Cerebellar abscess	Headaches, ataxia, cerebellar signs, neck stiffness etc.

incomplete penetrance. About half the patients have a positive family history. Its onset is typically in the second or third decades with a slightly higher incidence in females. The condition tends to be precipitated or promoted by pregnancy.

The association of osteogenesis imperfecta (Brittle bone disease), blue sclerae and otosclerosis is called Van der Hoeve's syndrome (Fig. 70).

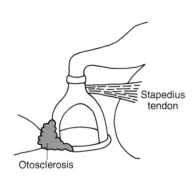

Fig. 69 Otosclerosis. New spongy bone is laid down in the region of the stapes footplate. This causes fixation of the stapes and conductive hearing loss. The ear drum is normal.

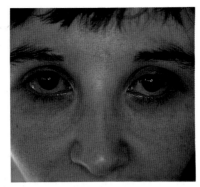

Fig. 70 Van der Hoeve's syndrome. This describes the association between osteogenesis imperfecta, blue sclerae, and otosclerosis.

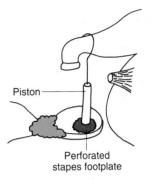

Fig. 71 Stapedectomy. The arch of the stapes is removed and is replaced by a piston thus restoring the mobility of the ossicular chain.

Table 5. Treatment options in otosclerosis

No treatment	This is often the best option especially when the hearing loss is slight or unilateral.
Hearing aid	Hearing aids work very effectively in conductive losses, especially when inner ear function is normal. This is because all that is required is amplification. Patients with otosclerosis should be encouraged to try a hearing aid, prior to embarking on surgery.
Stapedectomy	This operation involves removing the stapes and replacing it with a piston (Fig. 71). The operation results in a dramatic and prolonged hearing gain. It carries a small but definite risk of sensorineural hearing loss.

Diagnosis

The presence of a *normal ear drum* in a patient with *conductive hearing* loss usually implies otosclerosis.

Treatment

Options for treatment are outlined in Table 5.

Key Point

A normal ear drum in a patient with conductive hearing loss suggests otosclerosis.

CHAPTER SIX

Disorders of the inner ear

Inner ear disorders manifest themselves with sensorineural hearing loss, tinnitus, vertigo, or facial paralysis.

ACOUSTIC NEUROMA

These are benign tumours which arise from the auditory nerve. The tumours arise in the internal auditory canal (Fig. 72). Later, they expand causing cranial nerve palsies, brain stem compression and raised intracranial pressure. Patients with Von Recklinghausen's disease may have bilateral tumours (Figs 73 and 74).

The earliest symptom is *unilateral hearing loss* or *tinnitus*. Sometimes the hearing loss can be of sudden onset. Vertigo is uncommon. Patients with large tumours experience headaches, visual disturbance and ataxia.

Early diagnosis is crucial as the mortality and morbidity from surgery is directly related to tumour size. For instance, small tumours can be removed with preservation of the facial nerve (and sometimes hearing). The removal of large tumours can compromise the blood supply to the brain stem and preservation of the facial nerve is possible only rarely. The complications are thus formidable.

Most patients with acoustic neuromas in the UK present with neurological symptoms (i.e. large tumours) rather than with hearing loss. A major factor is 'doctor delay' in referring these patients for assessment.

Investigations

Pure tone audiometry will usually confirm unilateral or asymmetrical sensorineural hearing loss. Electric response audiometry (Fig. 75) is an essential screening tool to differentiate cochlear deafness from deafness due to tumours on the auditory nerve (retrocochlear disease). A caloric test will reveal an absent or depressed response. CT scanning (Fig. 76) and magnetic resonance imaging (Fig. 77) are the definitive diagnostic modalities.

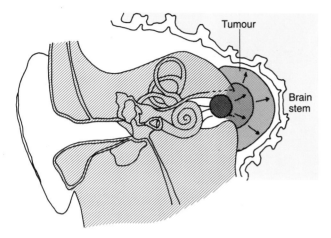

Fig. 72 The growth of acoustic neuromas. These benign tumours arise within the internal auditory canal (intracanalicular) and expand to compress the brain stem.

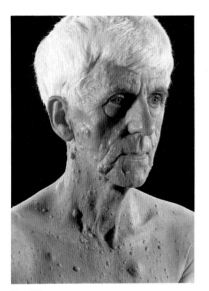

Fig. 73

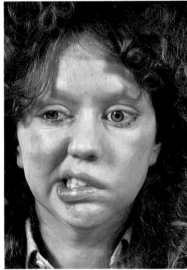

Fig. 74

Fig. 73 Von Recklinghausen's disease. Note the typical cutaneous neurofibromata. The patient presented with left sided hearing loss.

Fig. 74 Von Recklinghausen's disease. This 22-year-old girl had *bilateral* acoustic tumours. The tumour on the left side was removed leaving her with a facial paralysis and a 'dead' ear. The acoustic neuroma in the right ear will be allowed to remain as long as her hearing lasts, provided her neurological condition remains stable.

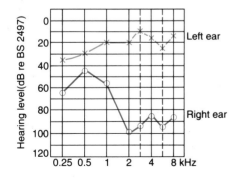

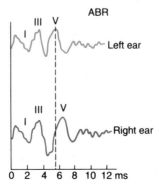

Fig. 75 Pure tone audiogram and auditory brainstem responses in a patient with a right sided acoustic neuroma. The hearing is normal in the left ear as is the I–V latency (broken vertical line represents upper limit of normal). On the right side, there is high frequency hearing loss and the I–V latency is prolonged. (*Note* Wave I—distal cochlear nerve, Wave III—cochlear nuclei, and Wave V—tracts and nuclei of the lateral lemniscus).

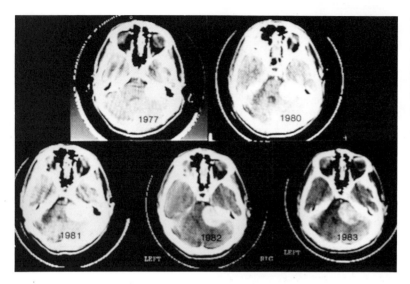

Fig. 77 Acoustic neuroma (4 mm tumour)—magnetic resonance imaging. This technology is superior to CT scanning at outlining neuromas and adjacent neural structures. Its limitation is that the temporal bone is not imaged.

Fig. 76 Acoustic neuroma—CT scans. These scans demonstrate the growth of an acoustic neuroma over a six year period in a patient who refused surgery.

Treatment

Surgical removal is the only option. Best results are achieved by a joint neurosurgical-otological approach. Some surgeons favour tumour removal through the ear (translabyrinthine approach), while others favour removal through the posterior fossa (sub-occipital approach). Physiological monitoring (Fig. 78) of the facial and auditory nerves during surgery has greatly improved the functional outcome.

> **Key Point**
> Patients with unilateral or asymmetrical sensorineural hearing loss should always be screened for acoustic neuroma.

Fig. 78 This patient, seen post-operatively, had a left sided acoustic neuroma removed, helped by physiological monitoring. Her facial nerve function and hearing were preserved.

PRESBYACUSIS

This is the term applied to hearing loss due to ageing. It is usually bilateral and symmetrical (Fig. 79). The age of onset is variable.

Audiometry often reveals a high frequency loss. The high frequencies are crucial for speech intelligibility. Consonants ('d', 'n', 't', 'l', 's'), which act as the 'stop gaps' in speech are generally in the high frequency range (Fig. 80). Vowels tend to be low frequency and are less important for understanding speech (Table 6).

Many patients with presbyacusis become unduly sensitive to certain sounds. This is because the cochlea loses its ability to accommodate its normally vast intensity range (over one million to one!) resulting in

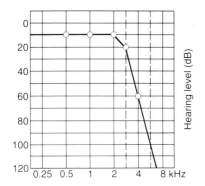

Fig. 79 Presbyacusis. The typical audiogram shows high frequency bilateral symmetrical sensorineural hearing loss.

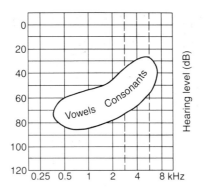

Fig. 80 The representation of vowels and consonants on a pure tone audiogram. Consonants are more important for understanding speech. Many patients with high frequency loss can hear the spoken voice but cannot make out what is being said.

Table 6. The relative importance of vowels and consonants. By removing consonants (high frequency) from the sentence, no intelligible message remains. Removing vowels (low frequency) has much less effect

Example:	*'The cat sat on the bag'*
Without consonants	*e a a o e a*
Without vowels	*th c t s t n th b g*

abnormal growth of loudness (called *recruitment*). This explains why shouting at a patient with a high frequency sensorineural loss often adds nothing to what they understand and may make matters worse.

Thus, many of these patients complain that while they know people are speaking they cannot understand what is being said. They often say that words merge one into the other or that the speech is 'muffled'. They function well on a 'one to one' basis but have great difficulty with group conversation and when there is background noise.

Management

Many patients are concerned that they may lose their hearing completely and reassurance on this front is important. Emphasizing their good low and mid frequency hearing is good for morale. It is also very useful to explain to patients and their relatives the special problems caused by high frequency hearing loss as this makes all parties more tolerant. Many, however, will need a properly fitted hearing aid which for most patients will offer substantial hearing improvement (see Chapter 9).

Key Point

Ageing affects both sides of the body! It does not typically cause unilateral hearing loss (cf. acoustic neuroma).

SUDDEN SENSORINEURAL HEARING LOSS

This is a medical emergency and patients should have the benefit of urgent ENT assessment. The hearing loss may be associated with a constitutional upset, often attributed to viral illness. The more severe cases tend to be accompanied by vertigo.

Management—early

Bed rest is important initially; *vasodilators*: the choice is wide. Carbogen gas (a mixture of 5 per cent CO_2 and 95 per cent O_2) is advocated by some to improve oxygen delivery to the damaged cochlea but this therapy needs hospitalization. Nicotinic acid, cinnarizine, beta histine hydrochloride all have their protagonists. *Steroids* are often given in a reducing dose over 10 days.

Management—late

It is essential to exclude an acoustic neuroma or other disease (e.g. syphilitic disease) in the auditory system.

Prognosis

This needs to be guarded. In general, low frequency losses recover better than high frequency deficits and severe vertigo is an unfavourable factor. Recovery usually takes place within 2 to 4 weeks but there are some late recoveries.

MENIERE'S DISEASE

This clinical condition, which tends to be over diagnosed, is characterized by intermittent attacks of vertigo, fluctuant hearing loss, and tinnitus. Very often, there is a sensation of pressure in the affected ear which may warn the patient of an impending attack. The hearing loss affects the low frequencies initially and is sensorineural in type. Later, the other frequencies may be affected. Occasionally, the condition may be bilateral. The vertigo usually lasts between 10 minutes and 8 hours and is often accompanied by nausea and vomiting. The condition most commonly affects adults (women more than men) between the ages of 35 and 55.

The cause of this condition is unknown but there is no shortage of ideas. Among the possible mechanisms are immunological, vascular, endocrine, hereditary, and psychological factors.

Pathology

Excessive accumulation of endolymphatic fluid (hydrops) is the most commonly observed finding. The distension may result in rupture of the inner ear membranes and mixing of endolymph (which is rich in potassium) and perilymph (which is low in potassium). This is the basis for the abrupt cochleovestibular failure which characterizes this condition.

Investigations

Pure tone audiometry. Evoked response audiometry is useful in excluding a retrocochlear cause (i.e. an acoustic neuroma) and electrocochleography (Fig. 81) may sometimes be diagnostic. Serological testing to exclude syphilitic ear disease should be performed. CT scanning and MRI may be necessary in selected cases to exclude acoustic neuroma.

Fig. 81 A normal electrocochleogram (above). Note the changes typical of an excess of fluid in the endolymph compartment (endolymphatic hydrops), as occurs in Meniere's disease.

Treatment

The natural history of Meniere's disease is decidedly towards spontaneous resolution. Reassurance is therefore extremely important.

Medical: acute phase. Bed rest, vestibular sedatives (diazepam is best), antiemetics (phenothiazines).

Prophylaxis. This is an area of some controversy. Betahistine has recently been shown to suppress histamine receptors in the vestibular nuclei and may thus help control troublesome dysequilibrium; it may also improve the cochlear microcirculation. Salt restriction also has its advocates.

Surgery. This is reserved for medical failures and thus has only a limited place in the treatment of Meniere's disease. Hearing conservation procedures (decompression of the endolymphatic sac, division of the vestibular nerve) are preferred to those procedures which destroy inner ear function (total labyrinthectomy).

Key Points

1. The diagnosis of Meniere's disease requires a triad of symptoms (fluctuant hearing loss, tinnitus, and vertigo). All that spins is not Meniere's disease!

2. Spontaneous resolution occurs in over 70% of patients with Meniere's disease.

3. Consider the diagnosis of acoustic neuroma in all patients with unilateral inner ear symptoms.

ACUTE VESTIBULAR FAILURE (VESTIBULAR NEURONITIS)

In this condition, there is a sudden, severe onset of incapacitating vertigo in a previously healthy patient. The patients are usually aged between 20 to 40. An important feature is the absence of hearing loss or tinnitus. The duration of the acute attack varies between 2 and 5 days, with mild dysequilibrium persisting over several weeks (Table 7). Examination confirms the presence of spontaneous nystagmus. Treatment is symptomatic (vestibular sedatives, especially diazepam in the initial stage). The cause is unknown but is often considered to be viral and the prognosis is excellent.

BENIGN POSITIONAL VERTIGO (BPV)

This condition is characterized by intermittent attacks of vertigo on adopting a sudden change in posture, i.e. on lying down, stooping, etc. It often follows head injury. The attacks last a few seconds and are not associated with any auditory symptoms (Table 7). Positional testing (Figs 82 and 83) is pathognomonic of this condition. Treatment is symptomatic and the prognosis is excellent although recovery may take up to two years.

BAROTRAUMA AND PERILYMPH FISTULA

In the inner ear the two fluid chambers are separated by extremely delicate membranes. Sudden pressure changes, such as may occur during flying or diving (or even sneezing or straining), may rupture these membranes causing mixing of the biochemically different fluids in the inner ear. This results in any combination of sensorineural hearing loss, tinnitus, and vertigo (Table 7). The fistula test (p. 20) may be positive. Many leaks seal spontaneously on bed rest. Closure of the leak is possible surgically and this typically alleviates the vertigo.

Key Points

1. Sudden hearing loss, tinnitus, and vertigo following barotrauma may indicate a perilymph fistula.

2. In the clinical diagnosis of vertigo the two most important factors are the duration of attacks and the presence or absence of hearing loss.

Table 7. The differential diagnosis of peripheral vertigo

Diagnosis	Duration of vertigo	Hearing loss/tinnitus
Meniere's disease	10 min–10 hr	yes
Acute vestibular failure (vestibular neuronitis)	2–5 days	none
Benign positional vertigo	seconds	none
Perilymphatic fistula	months to years	yes
Psychogenic	years	none

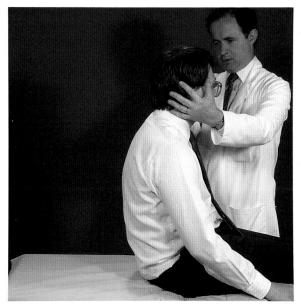

Fig. 82

Fig. 83

TRAUMA TO THE EAR AND SKULL BASE

Fractures of the temporal bone can be either longitudinal (80 per cent) or transverse (20 per cent). This is fortunate as transverse fractures cause more damage to the inner ear and facial nerve.

Examination

There are a number of important physical findings: haematoma over the mastoid bone (Battle's sign) (Fig. 84); blood in the external ear; laceration along the roof of the external ear; CSF otorrohoea or rhinor-rhoea (if the drum is intact, the CSF escapes down the Eustachian tube) (Figs 85 and 86); blood behind an intact ear drum (haemotympanum); conductive hearing loss from fluid in the middle ear or disruption of the ossicular chain) sensorineural hearing loss from cochlear fracture or concussion. The timing of onset of a facial paralysis (Fig. 87) is important to ascertain.

Treatment

General care of a patient with a head injury is necessary. Facial paralysis may call for exploration of the facial nerve and microneural repair (Fig. 88). Conductive hearing loss may require ossiculoplasty at a later stage. CSF rhinorrhoea will usually settle spontaneously but may need surgical repair. Persistent vertigo or imbalance may point to rupture of either the round or oval windows.

Fig. 82 Positional testing. The starting position of the patient on the couch must be such as to allow the head to be dropped below the end of the bed when the patient is thrust into a horizontal position (Fig. 83).

Fig. 83 Positional testing. Note that the head is below the level of the couch and rotated to one side. The patient's eyes should be open to observe nystagmus. With the affected ear lowermost, a rotatory nystagmus is produced after a latent period of a second or so. Repeated testing produces a progressively smaller response (fatigueability).

Key Points

1. Total facial paralysis immediately following head injury suggests major injury to the nerve. Delayed paralysis usually recovers spontaneously.

2. The ear canal should not be syringed in patients with temporal bone fractures.

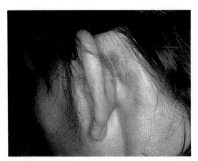

Fig. 84

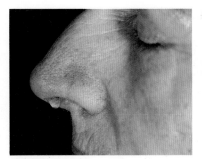

Fig. 85

Fig. 86

Fig. 84 Battle's sign. A haematoma over the mastoid is sometimes seen following temporal bone fracture.

Fig. 85 CSF Rhinorrhoea. This patient was referred because of a long standing 'runny nose'. Twelve years previously she had sustained a skull base fracture. Note the drip on the nasal tip. The unilateral rhinorrhoea had been attributed mistakenly to sinusitis!

Fig. 86 CSF Rhinorrhoea. The same patient as in Fig. 85. After bending over for half an hour this amount of fluid was produced. The leak was repaired.

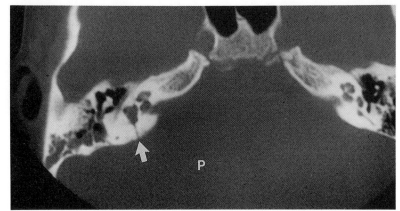

Fig. 87 Fracture temporal bone—CT scan. This fracture (arrowed) traversed the inner ear producing a 'dead ear' and total facial paralysis. A CSF otorrhoea followed due to a leakage from the posterior fossa (P).

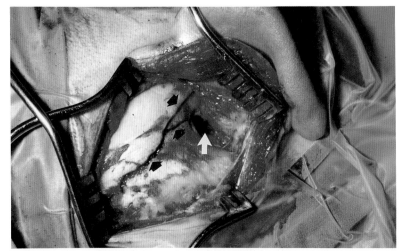

Fig. 88 Fracture temporal bone (right ear). The ear was explored to repair a damaged facial nerve. The fracture (black arrows) traversed the bony external ear canal (white arrow) on to the squamous portion of the temporal bone.

NOISE INDUCED HEARING LOSS

The inner ear can be damaged by sudden acoustic trauma (i.e. blast injury, gun-fire, etc.) or by prolonged exposure to excessive noise. In the acute injury, the sensorineural hearing impairment is greatest at very high frequencies and is often accompanied by tinnitus. With prolonged exposure, as in heavy industry, the hearing loss may be reversible initially. This is due to cochlear fatigue and is called a *temporary threshold shift*. This usually occurs within 2 hours of exposure. With further exposure, a permanent threshold shift occurs. The typical audiogram shows a notch at 4 kHz (Fig. 89) with gradual involvement of the lower frequencies with continued exposure.

The law requires that workers are protected from noise and strict criteria for permissible levels have been laid down. The damaging effects of noise may be minimized in a number of ways: by reducing the noise output from machines, by wearing ear defenders (cotton wool plugs do not protect from noise!), by keeping to a minimum the duration of exposure to the noise, by offering rest intervals and by having regular screening audiometry of personnel at risk.

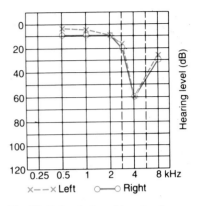

Fig. 89 Noise induced hearing loss. The external ear canal acts as a tube which resonates at 4 kHz. Hence the dip at this frequency following noise exposure.

Key Points

1. To attribute hearing loss to noise requires the patient to have had a genuine history of noise exposure. If the loss is not symmetrical, investigate for acoustic neuroma.

2. Cotton wool plugs may keep dust out of ear canals—they do not protect from noise.

AIDS AND THE EAR

The acquired immune deficiency syndrome (AIDS) can affect the ear. The major manifestations of this disorder are summarized in Table 8.

Table 8. Acquired immune deficiency syndrome and the ear

External ear	Kaposi's sarcoma, fungal otitis externa, necrotizing ('malignant') otitis
Middle ear	Acute and serous otitis media. Mastoiditis. Most common in paediatric AIDS
Inner ear	Sensorineural hearing loss (neuropathy of auditory nerve) Iatrogenic (vincristine, antifungal agents) Hearing loss progresses with disease Exclude neurosyphilis

OTOTOXICITY

The principal groups of drugs injurious to the inner ear are the aminoglycosides, diuretics (e.g. frusemide), salicylates, and chemotherapeutic agents. Some drugs selectively damage the cochlea (neomycin, kanamycin) while others damage the vestibular system (streptomycin). Gentamicin damages both systems.

Tinnitus is usually the first symptom, followed by progressive sensorineural hearing loss and vertigo.

Prevention is crucially important as there is little one can do to reverse the damaging effects of these drugs. Great care must be taken in patients with compromised renal function. Monitoring serum levels during treatment is essential. Serial audiometry is helpful. The use of non-ototoxic alternatives (i.e. cephalosporins) should always be considered.

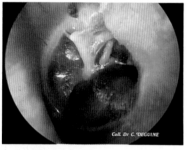

Fig. 90 Glomus jugulare. These benign tumours arise from the jugular bulb giving rise to the 'setting sun' sign shown here.

GLOMUS TUMOURS

These rare benign tumours arise from non-chromaffin paraganglionic tissue which has a wide distribution in the head and neck. In the neck, these cells give rise to carotid body tumours (p. 197). In the temporal bone, three types of tumour may arise depending on location: glomus tympanicum (arising in the middle ear), glomus jugulare (arising in the jugular bulb), and glomus vagale (arising near the point of exit of the vagus nerve from the temporal bone).

Symptoms

Tinnitus synchronous with the pulse beat is the classic symptom. Hearing loss (conductive or sensorineural), facial paralysis and paralysis of cranial nerves IX to XII may occur.

Examination

A pulsatile mass behind the ear drum ('setting sun' sign) (Fig. 90) is usually apparent. Sometimes an audible bruit over the temporal bone, facial paralysis or paralysis of cranial nerves IX to XII may be present.

Treatment

Surgery, radiation therapy, or a combination of both.

HERPES ZOSTER OTICUS (RAMSAY HUNT SYNDROME)

The herpes zoster virus may attack the spiral or vestibular ganglion in the inner ear or the ganglia of the facial nerve. The first symptom is intense pain in the ear (with little to be found on examination), followed a few days later by a vesicular eruption on the pinna and external ear. Sensorineural deafness, vertigo and facial paralysis may ensue (Ramsay Hunt syndrome) (Figs 91 and 92). Early treatment with the antiviral agent acyclovir is said to improve prognosis and reduce the likelihood of post herpetic neuralgia.

TINNITUS

Tinnitus is an hallucination of noises in the head or ears. It is a description of a symptom and not a diagnosis. It may occur on its own or be associated with hearing loss and vertigo. In certain unfortunate patients it can prove to be almost unbearable and drive them to distraction. Some such patients have committed suicide.

Management

These patients need careful neurotological assessment, especially those with unilateral symptoms (bear in mind Meniere's disease, acoustic neuroma, glomus jugulare tumours, intracranial vascular abnormality).

Once underlying disease is excluded, patient counselling is perhaps the most important aspect of management. Patients should be encouraged to manage the tinnitus within their own resources.

For those with hearing impairment, a hearing aid may be invaluable (by hearing better, the patient 'ignores' the tinnitus). If getting off to sleep is a problem, using the 'snooze' facility on a clock radio may be helpful. In others, night-time sedation may be required. *Tinnitus maskers* (p. 75) are also an important adjunct to therapy. Self help and support groups are useful and help to maintain morale.

NON-ORGANIC HEARING LOSS

It is important that this is recognized as it often represents a pitfall for the unwary. It is most likely to occur in the pursuit of compensation following head injury or alleged injurious exposure to noise. Other patients may feign total deafness to fulfil some psychological need (adolescent girls most commonly). Diagnosis depends on awareness and a number of audiological tests have been developed to support clinical suspicions. Electric response audiometry will confirm the diagnosis.

Fig. 91 Herpes zoster oticus—Ramsay Hunt syndrome. The patient developed ear ache followed by a vesicular eruption and facial paralysis (see Fig. 92).

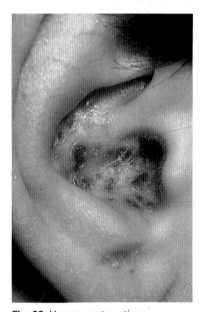

Fig. 92 Herpes zoster oticus—Ramsay Hunt syndrome. The characteristic vesicular eruption.

CHAPTER SEVEN

The facial nerve

....

The cell bodies of the facial nerve lie in the facial nuclei of the brain stem. The nerve leaves the brain stem in the cerebellopontine angle, directing its way towards the internal auditory canal. It then enters the temporal bone and has a complex Z-shaped course through it. Its course within bone is the longest of any nerve in the body—about 30 mm. Having left the temporal bone it then enters the parotid gland and fans out to supply the muscles of facial expression.

The cerebellopontine angle, temporal bone, and parotid gland are havens of pathology. Thus it is useful to consider lower motor neurone facial paralysis under the headings: intracranial, intratemporal, and extratemporal (Table 9).

Table 9. Causes of lower motor neurone paralysis

Intracranial	Intratemporal	Extratemporal
Meningioma	Acute and chronic ear disease	Parotid malignancy
Congenital cholesteatoma	Glomus tumours	Facial lacerations
Acoustic neuroma	Herpes zoster	
	Fracture	
	Post surgery	
	Temporal bone tumour	
	Bell's palsy	

Patients with facial paralysis need careful history-taking, ENT, and neurological examination. Always examine the parotid gland— remember it has a deep lobe which may displace a tonsil and can be seen and felt through the mouth. Bilateral paralysis (Figs 93 and 94) or partial paralysis (Fig. 95) can easily be overlooked.

In *investigations*, the degree of paralysis should be recorded. Photography of facial movements is an excellent method of recording the paralysis. Audiometry and measurements of stapedial reflexes are necessary in most cases. Electroneuronography (p. 24) can evaluate the severity of nerve injury and give a guide to progress and prognosis. Magnetic resonance imaging and CT scanning are indicated in selected patients.

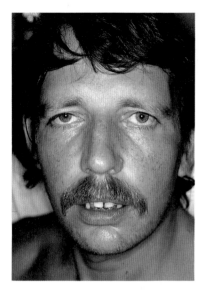

Fig. 93 Bilateral facial paralysis. At rest, no abnormality is apparent and the diagnosis can be overlooked. This patient had a head injury.

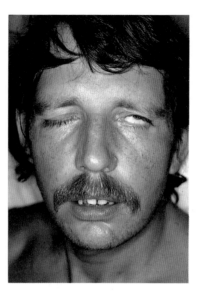

Fig. 94 Bilateral facial paralysis (the same patient as in Fig. 93). On forced eye closure, the weakness is readily apparent.

Fig. 95 Partial facial paralysis. The patient has a mild right facial weakness. Note the asymmetry of her smile.

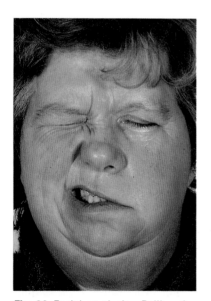

Fig. 96 Facial paralysis—Bell's palsy. The patient had a sudden onset dense left facial paralysis. Nerve function recovered fully in three months.

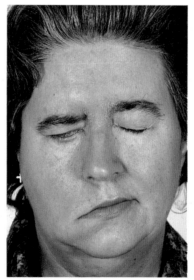

Fig. 97 Facial paralysis—'Bell's palsy'. The patient had a gradual onset facial paralysis which was erroneously diagnosed as Bell's palsy. She in fact had a malignant parotid tumour.

Fig. 98 Facial paralysis—'Bell's palsy'. Another paralysis incorrectly labelled 'Bell's Palsy'. The patient had a partial paralysis due to a tumour at the petrous apex.

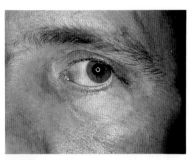

Fig. 99 Lateral tarsorrhaphy. Protection of the cornea is the top priority in the management of facial paralysis. The patient underwent a temporary tarsorrhaphy while awaiting recovery of his facial nerve.

BELL'S PALSY

The diagnosis of Bell's palsy should only be applied when no specific cause has been found for a facial paralysis after thorough investigation (Figs 96–98). Typically the paralysis is of sudden onset and is complete within 24 hours. Signs of recovery begin to appear within two months of the onset. Steroids are probably useful if given early in the course of the condition.

Care of the eye is the most important consideration in the early management of facial paralysis (Fig. 99). Without a blink reflex, the cornea is unprotected and may ulcerate. This is particularly so if corneal sensation is impaired (note that the trigeminal and facial nerves are close together at the petrous apex). If in any doubt, early referral to ophthalmology is essential.

It follows that partial paralysis of gradual onset should not be labelled 'Bell's palsy'. Investigations in such a patient are likely to reveal a specific cause.

Recovery

This depends on the severity of nerve injury. If Wallerian degeneration takes place, recovery is likely to be delayed and incomplete. Overall, 90 per cent of patients have satisfactory return of facial function. Surgery to decompress the nerve is necessary only in exceptional cases.

Key Points

1. All that palsies is not Bell's! The diagnosis of Bell's palsy can only be made by exclusion of other causes of facial paralysis.

2. Protection of the cornea is the most crucial factor in the early management of facial paralysis.

3. Beware of a malignant parotid tumour (remember the deep lobe!) as a cause of facial paralysis.

SURGERY OF THE FACIAL NERVE

Key Points

1. The entire facial nerve is amenable to surgical repair.

2. Many valuable techniques exist to reanimate a paralysed face when the nerve cannot be repaired.

The entire facial nerve, from brain stem to the face, is amenable to surgical repair (Figs 100 and 101). This may involve the combined efforts of an otologist and a neurosurgeon. Microsurgical techniques allow direct 'end to end' repair or the interposition of a nerve graft.

Patients whose nerves cannot be repaired in this way can benefit from other techniques that re-animate the paralysed face. These can improve the patient's appearance, speech, and morale. Patients should be referred to otologists doing such work.

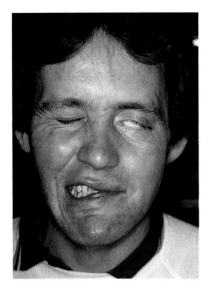 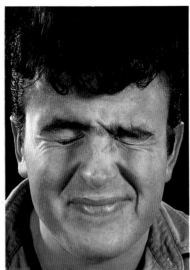

Fig. 100 Late facial paralysis. The patient sustained a fractured temporal bone two years previously. This caused a 'dead' ear and total left facial paralysis. No attempt was made to repair the damaged nerve. The result is disappointing (see Fig. 101).

Fig. 101 Facial nerve surgery. The patient had a dense right facial paralysis following temporal bone fracture. The nerve was explored and a fragment of bone was removed from the nerve. There was almost complete recovery of facial nerve function.

CHAPTER EIGHT

Paediatric otology

The ear diseases described in previous chapters can all affect children. However, certain conditions occur rather more frequently in children. Hearing impairment also has a different, and possibly greater, impact on children than it does on adults. Speech and language are dependent on good hearing during the critical period of their acquisition. Hearing impairment can also cause developmental delay of motor and social skills and may have profound effects on educational achievement. Hearing is also fundamental to the normal rapport a child has with parents, siblings, and other children. There is thus a sense of urgency about identifying and treating hearing loss in infants and children.

EXAMINATION OF THE EAR

Ear examination has been described in Chapter 2. It is essential for the conduct of the examination to have control over both the otoscope and the child. If the child is fractious, as may happen in otitis media, ask the mother or a nurse to hold the child as shown (Fig. 102). If often helps to let the child see the otoscope and to shine the light on to his hand or face before introducing it to the ear. The otoscope should be held like a pen with the examiner's finger on the child's cheek acting as an 'early warning system' lest the child should jolt. The pinna should be gently retracted backwards to straighten the S-shaped canal. If wax obscures the view, it can be syringed or removed with a probe. An instrument must never be introduced into a child's ear canal until the situation is entirely under control. It is better to defer examination rather than risk injuring a child's ear or shattering its confidence.

Fig. 102 Examination of the ear in children. The parent (or nurse) should restrain the child's upper limbs. The lower limbs can be controlled by locking them between the thighs (failure to do so can result in an embarrassing injury to the examiner!).

AUDIOLOGICAL ASSESSMENT

The most important rule to remember is this: if parents suspect their child has hearing impairment—*believe them!* The younger the child, the more important it is to take stock. Always arrange a formal assessment and, if needs be, repeat assessments if doubt continues.

Screening at birth

All 'at risk' children should be screened shortly after birth. The main 'at risk' determinants are as shown in Table 10.

Attempts have been made to screen infants at risk by using the fact that sound normally causes changes in heart rate or respiration ('crib-o-gram', auditory response cradle, etc.) but these methods are not sufficiently reliable. More effective screening involves electric response audiometry (p. 22) but this is not widely available and also has certain limitations, especially when testing the responses of a relatively immature nervous system. Acoustic emission testing (p. 24) may occupy an important place in screening in the future.

Table 10. 'At risk' factors for sensorineural hearing loss in childhood

Family history of deafness,
Maternal infections: rubella, toxoplasmosis, syphilis, viral infection (cytomegalovirus)
Pre-maturity
Other congenital abnormalities
Anoxia
Jaundice
Meningitis
Ototoxics (e.g. gentamicin)

Parental suspicion

The importance of heeding parental suspicion cannot be overemphasized. In this context, the myth of the 'fussy mother' can safely be buried. In some centres, guidelines are given to every parent of a new-born child. These indicate how babies and infants with normal hearing react to sound. This heightens parental awareness and facilitates early diagnosis, especially in children with no risk factors. This approach cannot be relied upon completely as some hearing losses, especially those in the high frequencies, are likely to be overlooked.

Health visitor

For most children, the first hearing check is performed by the health visitor at between 6 and 8 months of age. The technique used is called *distraction testing*. This requires a co-operative child and considerable skill on the part of the tester. The child usually sits at a table on his parent's lap and his attention is obtained by an observer (Figs 103 and 104). The tester introduces a sound surreptitiously from behind the child. The child's reaction is noted by the observer. The results of screening depend on the enthusiasm and training of health visitors and the facilities available to them to do these assessments. Alas, many high risk children (ethnic minorities, low socio-economic groups) do not attend these clinics or will not bring their children along for definitive testing.

Conditioning audiometry

From age 18 months to 3½ years a variety of techniques exploit the ability of children of different mental ages to undertake specific tasks. The intensity and frequency of the stimulus can be altered. Given a co-operative child and a skilled tester, a reliable assessment of hearing acuity and discrimination can be obtained.

From about the age of 4 years a pure tone audiogram can be obtained.

Electric response audiometry (ERA)

Children with intellectual impairment or behavioural problems may not allow testing by conventional methods. In these children electric response audiometry (see p. 22) may be necessary to determine hearing levels (Fig. 105). Occasionally it may be necessary in otherwise normal children where doubt persists after conventional testing. ERA usually requires the child to be either asleep or sedated.

ACUTE SUPPURATIVE OTITIS MEDIA

This is an acute inflammatory process characterized by purulent fluid in the middle ear. It is one of the most common childhood infections. The reasons are poorly understood but are due, at least in part, to the relative immaturity of the Eustachian tube and immune defence mechanisms.

Symptoms

Irritability, progressive throbbing pain, fever, and hearing impairment are the usual symptoms. Untreated, copious purulent discharge from the ear will follow when the ear drum perforates and at this point the pain subsides.

Examination

Be careful—a child with otitis media is usually fractious and will allow only a fleeting glimpse down its ear canal. Sometimes the diagnosis will have to be presumptive. Position the child securely and ensure full control of the otoscope. Inspect behind the pinna for swelling over the mastoid bone (occurs in mastoiditis). The drum will be seen to be hyperaemic (Fig. 106) and often bulging due to pressure from accumulated fluid in the middle ear.

Treatment

Amoxicillin or ampicillin are the preferred initial drugs (the most frequent infecting organisms are *Streptococcus pneumoniae* and *Haemophilus influenzae*) and should be given for 10 days. If the child is allergic to penicillin, a trimethoprim–sulphamethoxazole combination

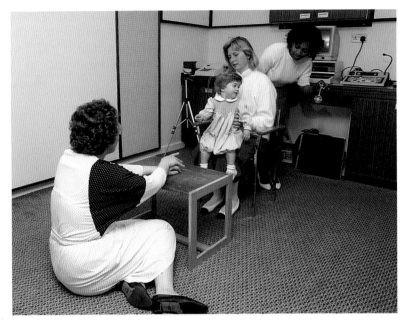

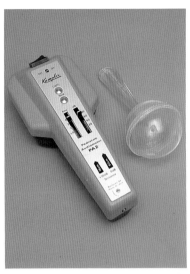

Fig. 104 Warble tone generator and Manchester rattle. These are frequently used in distraction testing.

Fig. 103 Distraction testing. When the child's attention has been obtained by an observer, a sound is skilfully introduced by the tester. No visual clues must be given to the child. There should be a definite rotation of the child's head towards the sound source.

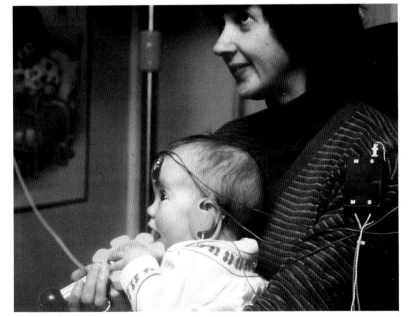

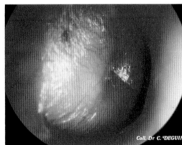

Fig. 106 Acute otitis media. The tympanic membrane is bulging, red, and injected with many tiny blood vessels.

Fig. 105 Electric response audiometry. Sound stimuli are delivered through a headphone as the child sits on its mother's lap. Recording electrodes are placed on the scalp.

can be used. Paracetamol is often indicated for pain relief and to lower body temperature. Myringotomy to drain the fluid is rarely done in the UK but is widely practised in the US. Its advocates find it useful to obtain samples for culture, especially in patients failing to respond to standard antibiotics.

> **Key Points**
>
> **1.** Pain in the ear does not always mean ear infection. Antibiotics should only be given when there is evidence of suppuration.
>
> **2.** Ear drops do not have analgesic properties. Oral analgesics are more likely to give pain relief.

RECURRENT OTITIS MEDIA

This can be a most frustrating condition for parents and clinician alike. Parental fatigue from many nights of disturbed sleep is coupled with their anxiety about repeated courses of antibiotics to which they say their child has become 'immune'.

A search for a nidus of infection elsewhere, e.g. sinusitis, is mandatory. An underlying allergy or immune deficiency may need exclusion. Not uncommonly these children have a persistent middle ear effusion which acts as a broth to culture further organisms.

Treatment

Prophylactic antibiotics are most helpful in this situation. This can be continuous low-dose amoxicillin or sulfisoxazole for a period of three to six months. Surgery, usually grommet insertion, can also be useful to treat an underlying persistent effusion especially if hearing impairment compounds the problem.

ACUTE MASTOIDITIS AND MASKED MASTOIDITIS

The incidence of this condition has greatly diminished with the widespread use of antibiotics for otitis media. Antibiotics have made the manner of presentation much more insidious and the danger now is that the diagnosis may go unrecognized. The term 'masked' mastoiditis has thus been coined.

Pain and swelling behind the ear are the typical findings. Commonly, the pinna may be pushed forward. The angle the pinna makes with the

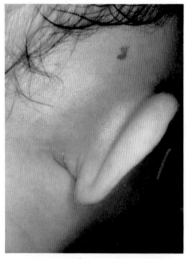

Fig. 107 Acute mastoiditis (child aged 6 months). The mastoid is red and swollen with the pinna pushed outwards.

side of the head may be widened (Fig. 107). The ear canal is swollen and it is often not possible to see the ear drum. A past history of ear discharge is a helpful indicator.

Untreated, there is a significant risk of intracranial complication (see p. 39).

Treatment

Requires hospitalization and involves intensive parenteral antibiotics while carefully monitoring the patient's progress. Mastoid exploration may be needed if the situation fails to resolve in 12–24 hours.

OTITIS MEDIA WITH EFFUSION (OME) (CHRONIC SEROUS OTITIS MEDIA, GLUE EAR)

This condition has reached epidemic proportions in children in Western countries.

Aetiology

Many factors have been implicated: social class, environment, immunity, allergy, Eustachian tube insufficiency, sinus infection and antibiotic treatment of suppurative otitis media. Oxygen is continually being absorbed by the respiratory epithelium resulting in negative pressure in the middle ear. This encourages transudation of fluid from the mucosa.

Symptoms

The highest incidence is in the age group 2 to 5 years. The various modes of presentation may be summarized as follows:

1. *Hearing impairment.* This may fluctuate depending on the volume of fluid in the middle ear.
2. *Language delay.* The peak incidence of this condition is at the 'critical period' for speech and language development.
3. *Behavioural problems.* Because a child cannot hear, he is likely to be labelled 'disobedient' or 'inattentive' both at home and at school. He may be described as being detached from his peers and 'being in a world of his own'.
4. *Recurrent ear infections.* The 'glue' is an ideal culture medium for micro-organisms.
5. *Reading/learning difficulties* at school.
6. *The 'silent syndrome'.* A child may be detected with florid serous otitis media purely on screening.

It is important to note that the condition produces hearing impairment at an important period in a child's development, i.e. when speech is being acquired and the child is about to begin school.

Examination

The otoscopic findings are of a dull, featureless immobile tympanic membrane (Figs 108 and 109). Sometimes fluid levels can be seen (Fig. 110). Often, the tympanic membrane is indrawn towards the middle ear (retracted) (Fig. 111).

Treatment

As in any condition where the cause is unknown, doctors differ and so do their treatments. An attempt is made in Table 11 to summarize the rationale and shortcomings of some common approaches to the problem. A hearing aid may be appropriate in certain cases.

It is important to bear in mind that about 50 per cent of effusions resolve spontaneously within six weeks of onset. In many instances, a 'wait and see' policy is appropriate.

Ventilating tubes (grommets)

Ventilating tubes (grommets) are tiny tubes inserted in the tympanic membrane (Fig. 112). Their purpose is to promote the ventilation of the middle ear rather than drainage of fluid. Insertion usually results in a dramatic gain in hearing and relief from earache. They remain in position for about 6 to 18 months and are extruded spontaneously by the tympanic membrane.

Ear discharge may follow grommet insertion. This is best treated by dry mopping the ear canal and inserting antibiotic/steroid ear drops. Occasionally systemic antibiotics are necessary. If discharge persists, refer to the ENT department.

Whether or not a child can go swimming is very much dependent on the individual surgeon. Some prohibit swimming completely. A number of studies, however, show that grommets do not increase the likelihood of infection after swimming. Thus, after a period of two weeks, some surgeons are happy for their patients to swim, even without ear plugs.

Some children require repeated grommet insertion (about 1 in 5 of children requiring surgical treatment) and parents tend to become disillusioned. In such patients, it is imperative that an underlying aggravating cause (e.g. allergy, sinus disease, adenoid hypertrophy) is sought and treated. Recurrent or persistent glue ear that defies the usual remedies can be extremely difficult to manage.

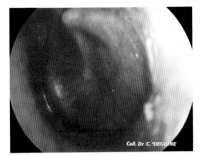

Fig. 108 Otitis media with effusion (glue ear). This dull, featureless tympanic membrane is very typical of glue ear. Compare with Fig. 109.

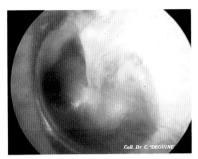

Fig. 109 A normal tympanic membrane.

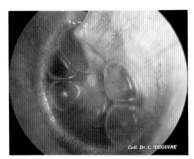

Fig. 110 Otitis media with effusion (glue ear). Air-fluid levels can be seen behind the tympanic membrane.

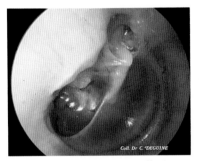

Fig. 111 Otitis media with effusion (glue ear). Often the tympanic membrane becomes retracted (indrawn) making the lateral process of the malleus more prominent. Note the dull appearance of the drum.

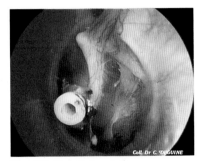

Fig. 112 Ventilating tubes (grommets). Note that the middle ear is well ventilated.

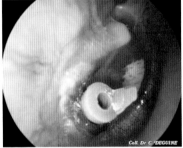

Fig. 113 Ventilating tubes (grommets) and tympanosclerosis. Following grommet insertion a small horse-shoe shaped ring of tympanosclerosis (white 'chalk patches') can form in the tympanic membrane.

Grommets may cause tympanosclerosis (Fig. 113) of the tympanic membrane but this is generally considered not to affect hearing. Rarely, a tiny perforation (Fig. 114) may persist after grommet extrusion.

Possible sequelae of otitis media with effusion are outlined in Table 12.

SENSORINEURAL HEARING LOSS

There are two main causes: *genetic* and *acquired* (prenatal or postnatal).

Genetic

This is overwhelmingly the most important single cause of sensorineural hearing impairment in childhood. Most (90 per cent) are due to autosomal recessive genes.

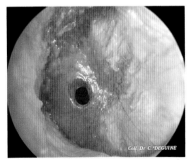

Fig. 114 Perforation following grommet extrusion. Occasionally a pin-hole perforation may persist after extrusion of the tube. This may need surgical repair.

Table 11. Treatment options for O.M.E. (otitis media with effusion—'glue' ear)

Treatment	Advantage	Disadvantage
No treatment	Natural history favours spontaneous resolution	May involve prolonged periods of hearing loss
Decongestants	Avoids hospitalization	No proven value. Can cause behavioural disorders. Expensive in long term
Long term antibiotics	Avoids hospitalization	Not a long term cure. Not all children benefit
Myringotomy alone	Low risk. Immediate hearing gain. Can be done as day case	Relief shortlived. Requires anaesthesia and hospitalization
Grommet insertion	As for myringotomy Relief from hearing loss 6–9 months	Requires anaesthesia and hospitalization Ear may discharge. Swimming forbidden by some surgeons. May scar ear drum (tympanosclerosis) or cause a perforation
Adenoidectomy	Improves Eustachian tube function. Relieves catarrhal symptoms	Needs anaesthesia and hospitalization. Not usually done as a day case. May bleed post-operatively. Benefit disputed.

Acquired

Prenatal. Certain infections acquired *in utero* can affect the developing auditory system (e.g. rubella, cytomegalovirus, syphilis).

Postnatal. Bacterial meningitis, which has a very high incidence in the first two years of life, is the most important cause. About 1 per cent of children with meningitis will have profound bilateral sensorineural hearing loss.

Management

Crucially important in management is early and accurate diagnosis. Only then can appropriate rehabilitative measures be instituted. The most important factor in rehabilitation is the early fitting of appropriate hearing aids which can be fitted as early as 6 months. Without hearing aids, the chance to enable severely deaf children to acquire speech and

Table 12. Possible sequelae of otitis media with effusion ('glue ear')

Atelectasis	A weak, floppy tympanic membrane which tends to be sucked into the middle ear. The drum may get stuck to the ossicles or to the promontory (adhesive otitis). Such thin drums readily perforate
Retraction pockets	With advancing atelectasis, the ear drum may form pockets which tend to collect epithelial debris
Erosion of ossicles	The long process of the incus tends to get eroded in atelectatic ears
Tympanosclerosis	Results from the drum developing calcific changes from repeated infection
Cholesteatoma	An uncertain association exists between chronic glue ear and cholesteatoma. This may be the result of retraction pocket formation or of metaplastic changes in the middle ear mucosa

language may be lost forever. Genetic counselling should always be offered to parents.

Education

Special arrangements for the child's education need to be made depending on the severity of the hearing loss. Some will attend ordinary school wearing their hearing aids. Those requiring more help will need the services of a Partial Hearing Unit attached to a main school. Special residential schools are necessary for those with total or near total deafness.

Sign language

There is much debate about what place sign language should occupy in the education of deaf children. On the one hand there are those who feel it should be the sole method of communication. Others feel that oral communication (i.e. lip reading) is the most valuable form of communication to teach a severely deaf·child. Many now feel that a combination of these methods (total communication) offers the profoundly deaf child the best chance of integration in the hearing world.

CONGENITAL DEFORMITIES OF THE EAR

'Bat' ears

Abnormally protruding ears can make a child the object of derision from his peers and cause emotional problems. Therefore, the deformity

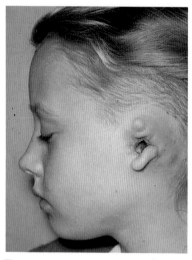

Fig. 115 Microtia. This child has had innumerable plastic surgical procedures to reconstruct the pinna. As is so often the case, the result is disappointing.

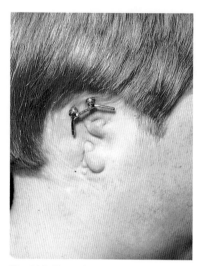

Fig. 116 Osseointegration. Titanium screws can be fixed in the mastoid bone to provide anchorage for a prosthetic pinna.

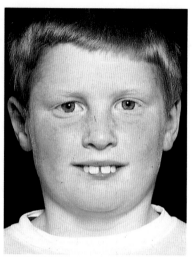

Fig. 117 Osseointegration. The end result (compare with Fig. 115). This technology saves the trauma and disappointment of conventional techniques of pinna reconstruction. A hearing aid can also be anchored in bone at the same time (Figs 123 and 124)

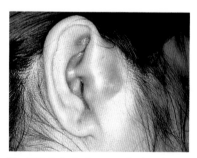

Fig. 118 Infected pre-auricular sinus. This is due to incomplete fusion of the hillocks that form the ear. Note the tiny pit which is always present.

ought to be corrected before school age, i.e. about the age of 5 years. The operation of pinnaplasty permanently 'pins back' the offending ears.

Congenital deformity of the pinna (microtia)

The severity of this deformity is variable. Sometimes the pinna may be absent or rudimentary. These deformities are often associated with an absent ear canal and middle ear. Fortunately, the inner ear is nearly always normal due to its totally different embryological development. Hence, it is essential that these children have their hearing assessed and hearing aids fitted at the earliest possible opportunity. The Treacher Collins syndrome is a combination of multiple facial deformities with abnormalities of the pinna and middle ear.

Treatment. The results of multiple plastic surgical procedures to reconstruct the pinnae on these children have been abysmal (Fig. 115). Children and their parents can now be spared this trauma thanks to advances in osseointegrated implants (Figs 116 and 117). Not only does this result in a superior cosmetic result but a bone-anchored hearing aid can also be implanted in bilateral cases (see Figs 123 and 124 on p. 73).

Pre-auricular sinuses

These are due to incomplete fusion of the primitive tubercles that form the pinna (Fig. 118). They can cause troublesome discharge and are often more extensive than at first appears. Treatment is by surgical removal.

CHAPTER NINE

Hearing aids, cochlear implants, and tinnitus maskers

HEARING AIDS

Over 200 000 hearing aids are issued by the National Health Service (NHS) in the UK annually. With an ageing population in most western countries, these figures are likely to rise even higher. Hearing aids function by selectively amplifying sound. They work best for conductive losses because inner ear function is usually normal and the problem is purely one of sound amplification. They are thus ideal in otosclerosis or stable chronic ear disease. In general, the better the inner ear function, the more efficiently the hearing aid will work. Thus, given a choice between two ears with hearing impairment, it is usually best to fit the aid on the 'good' ear. This often comes as a surprise to the patient!

Components

Miniature microphone, amplifier, receiver, and ear mould
The *microphone* picks up the incoming sound.
The *amplifier* makes the sound louder.
The *receiver* feeds the sound into the ear.
The *mould* is worn in the external ear.

Limitations

Hearing aids amplify sounds. This means that background noise (usually low frequency) is also amplified which can interfere significantly with speech intelligibility. Patients with good low-frequency but poor high-frequency hearing are most troubled by this. Some patients with impaired cochlear function are exquisitely sensitive to noise (a phenomenon called *recruitment*) and getting these patients to benefit from hearing aids may be difficult or impossible.

Some patients find hearing aids either cosmetically or socially unacceptable. They feel it draws attention to their disability which they wish to hide but modern miniaturized aids have done a lot to reduce this. Others feel hearing aids are associated with old age and might make them less competitive in the workplace.

Hearing aids need an insert (mould) in the ear canal and this can cause otitis externa. The precise fitting of an ear mould is of crucial importance and requires considerable care. Hearing aids cannot be worn in discharging ears.

In the future, better signal processing (peak clipping, automatic volume control, etc.) will permit more selectivity in the amplification and permit a better match between hearing aid and hearing loss.

Hearing therapy

It must be emphasized that the provision of a hearing aid marks the beginning of a rehabilitation programme. Alas, many patients reject their hearing aids because of inadequate counselling and support. Some deafened patients may be utterly dejected by their disability, especially when their employment is threatened. Enabling patients with multiple handicaps to use hearing aids successfully is a major challenge. Hearing therapists advise on lip reading, environmental aids, hearing aid skills and undertake auditory training; their services should be sought in appropriate cases.

HEARING AID TYPES

Behind the ear (BE)

These aids are the most commonly prescribed on the National Health Service (Figs 119 and 120). They can be quite powerful and are cosmetically acceptable. Adjusting the settings needs a certain amount of manual dexterity and can be a problem in the elderly, especially those with arthritis or neurological disability.

Body worn (BW)

These aids may be necessary for patients with profound hearing losses or those who do not have the manipulative skills to manage a 'behind the ear' aid. The microphone is on the device and the patient's clothing may impair the sound uptake. Friction between the clothes and the microphone can also be annoying.

'In the ear' and 'in the canal' aids

These aids (Figs 121 and 122) are accommodated entirely in the external ear. They are suitable for mild or moderate hearing losses. They are cosmetically more acceptable than the 'behind the ear' variety. They are, however, quite expensive.

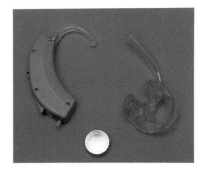

Fig. 119 'Behind the ear' (BE) aid assembly. The hearing aid is connected to a mould (on the right) which is inserted in the ear canal. The mould is individually made for each patient so as to obtain a comfortable fit. These aids are powered by small batteries (centre).

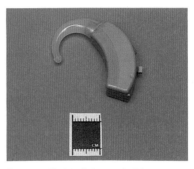

Fig. 120 'Behind the ear' aid— 'micro'. This tiny device can be fitted to babies a few months old.

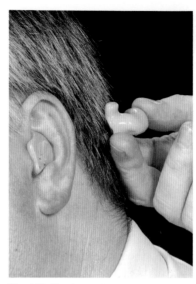 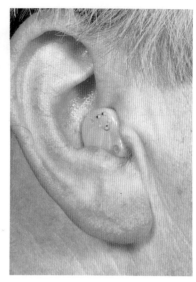

Fig. 121 'In the ear' aid. The attractiveness of these aids is readily apparent. The part that is inserted in the ear canal is shown on the right.

Fig. 122 'In the canal' aid. These ingenious devices fit in the ear canal. Regrettably, they are rather expensive.

Bone anchored aids

These have supplanted the conventional bone conducting aids which were fixed by means of a headband. The bone anchored aids are fixed by means of a special titanium screw in the mastoid bone and are highly efficient sound conductors (Figs 123 and 124). They are most useful in congenital ear abnormalities where there may be no pinna on which to hang an ordinary aid. They can also be worn by patients with discharging ears.

Implantable hearing aids

A recent advance has been the development of wholly implanted middle ear hearing aid systems. It is likely to be some years before these become widely available.

COMMERCIAL VERSUS NHS HEARING AIDS

In the UK, NHS aids are available free of charge. Commercial aids can be very costly (some cost over £700). The maintenance (repairs, batteries, etc.) of NHS aids is carried out without cost to the patient. Financial considerations are especially important to the elderly who may no

---

Key Points

1. It is better that a patient tries a hearing aid and rejects it than never to be offered one at all.

2. The provision of a hearing aid marks the beginning of a rehabilitation programme to improve communication skills.

3. Always fully examine the ears in a patient with hearing loss. Any suspicion of underlying disease requires referral to ENT (remember acoustic neuroma, etc.).

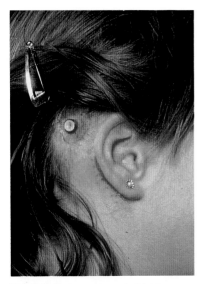 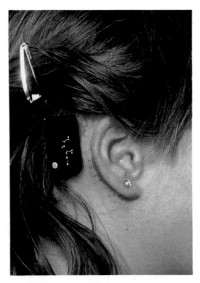

Fig. 123 Bone anchored aid. These are necessary when the patient has no pinna or is unable to wear conventional hearing aids (e.g. due to atresia of the ear canal). The first stage is the insertion of a titanium screw which becomes anchored solidly in bone.

Fig. 124 Bone anchored aid. The aid is clipped on the screw which transmits the sound vibration through bone to the cochlea.

longer be in employment and for whom the cost of a commercial aid may need to be met out of savings.

A wider selection of very sophisticated aids is available commercially, especially 'in the ear' varieties which are prescribed only for special medical reasons on the NHS.

Patients opting to buy hearing aids should be advised to see a reputable dealer and insist on a free trial of the device in their everyday environment. This may mean trying several aids. The patient should be advised against judging the value of an aid on its performance in the acoustically ideal environment of a sound treated room.

The greatest criticism of NHS aids is the time delay between referral for assessment and fitting of the device. Direct referral by family doctors to hearing aid clinics may help overcome the problem.

COCHLEAR IMPLANTS

Cochlear implants ('bionic ears') represent a breakthrough in the management of acquired total deafness. These patients derive no benefit from even the most powerful hearing aids. The deafness may be the result of meningitis, head injury, ototoxic drugs or some other cause.

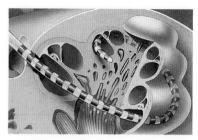

Fig. 125 (a) Cochlear implant. This is implanted at surgery. The tapered wire containing the electrodes leads off from the electronics package.

Fig. 125 (b) The electrodes are inserted into the spiral of the cochlea to stimulate the surviving auditory nerve endings.

Fig. 126 Cochlear implant. This child was totally deafened by meningitis. An ear-level microphone is worn like a hearing aid. The sound is converted to electrical signals in a body worn speech processor. The black circular device behind the ear is the transmitter coil which transmits signals to the implanted electronics.

The system consists of an ear level microphone which collects sound. A body-worn speech processor, about the size of a personal stereo, converts the sound to electrical signals. These signals are then passed to electronic circuitry implanted at the time of surgery. Tiny electrodes convey the signals directly to the auditory nerve (Figs 125 a and b), thus bypassing the function of the cochlea. In some systems, the electrodes remain outside the cochlea (extracochlear), while in others the cochlea is entered (intracochlear). Intracochlear devices afford better speech perception, with some patients being able to use a telephone. Cochlear implants have also been found to suppress tinnitus and can be used in young children (Fig. 126).

TINNITUS MASKERS

The rationale for this approach is to use an extraneous noise which drowns out or 'masks' the noise of the tinnitus. This gives the patient a sense of control over the symptom. It follows that the patient's tinnitus must be of such a quality as to allow this matching to take place. Tinnitus that varies in quality or is very bizarre may be impossible to mask. Maskers are worn like hearing aids (Fig. 127). They are not a 'cure' for tinnitus and should only be used as part of a comprehensive rehabilitation programme.

Fig. 127 Tinnitus masker. This device is worn in the ear canal and can produce a sound similar to the patient's tinnitus. This has the effect of masking or 'drowning' the indigenous noise.

PART TWO

The Nose and Sinuses

CHAPTER TEN

Clinical anatomy and physiology

ANATOMY OF THE NOSE

The external nose

The nasal skeleton consists of the nasal bones, the paired lateral cartilages, and the septal cartilage (Fig. 128). The proportion of the nose made up by the nasal bones varies between 30 and 70 per cent. Each nasal bone is attached to the frontal bone and the maxilla. Trauma to the nose can fracture the nasal bones which may be deviated or depressed.

There are paired upper and lower lateral (alar) cartilages. Each upper lateral cartilage is attached to the under-surface of the nasal bone and to the nasal septum. The septum, therefore, provides support to the dorsum of the nose. The delicate lower lateral cartilages are responsible for the support and appearance of the nasal tip. The shape of the nasal bones and cartilages has a bearing both on the appearance and the function of the nose.

The arterial supply to the external nose comes from branches of the facial and ophthalmic arteries. These vessels may bleed significantly after facial trauma. The angular vein lies at the medial canthus and it is via this vein that infection of the face can spread to the cavernous sinus. The angular vein may be damaged during rhinoplasty and this can result in bruising (black eyes). The lymphatic vessels of the external nose drain to the upper deep cervical chain.

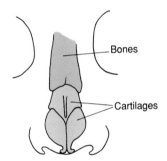

Fig. 128 The nasal skeleton consists of two nasal bones, two pairs of lateral cartilages (upper and lower), and the septal cartilage.

The nasal cavity

The septum divides the nasal cavity into two halves. The floor of the nasal cavity runs in a horizontal direction, parallel with the hard palate. The vestibule is the skin covered area at the entrance of each nostril. Hairs cover part of the vestibular skin so that a furuncle may arise in this region. The nasal valve is the narrowest portion of the nostril and demarcates the vestibule from the nasal cavity proper. The columella is the strut at the caudal end of the septum between the two nostrils.

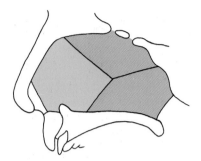

Fig. 129 The septum consists of the quadrilateral cartilage, and vomer and ethmoid bones.

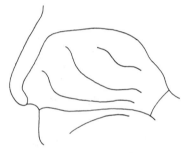

Fig. 130 The lateral wall of the nose.

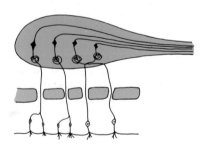

Fig. 131 The olfactory epithelium lines the superior part of the nose. The filaments of the olfactory nerve are vulnerable to injury as they pass through the cribriform plate.

The septum

The septum, which supports the dorsum of the nose, consists of the quadrilateral cartilage joined to the vomer and the ethmoid bones (Fig. 129).

The lateral wall of the nose

The lateral wall of the nose has three projecting shelves of bone known as turbinates or conchae (Fig. 130). These serve to increase the surface area of the nasal cavity. The turbinates are labelled superior, middle, and inferior and the space under each turbinate is called a meatus.

The middle meatus lies under the middle turbinate and is the most important functional area. All the sinuses open into this meatus with the exception of the sphenoid and posterior ethmoidal cells. The openings of the sinus ostia into the middle meatus are close together and form the *ostio-meatal complex*. This is a key area because pathology in this region can interfere with ventilation and mucociliary clearance of the sinuses.

Fibreoptic endoscopes allow the surgeon to inspect and operate on the ostio-meatal complex.

Key Points

1. All the sinuses with the exception of the posterior ethmoidal cells and the sphenoid sinus open into the middle meatus.

2. The ostiomeatal complex in the middle meatus is the key area for endoscopic sinus surgery.

3. Only one structure, the nasolacrimal duct, opens into the inferior meatus.

Epithelium lining of the nasal cavity

There are two types of epithelium within the nose: olfactory and respiratory.

Olfactory epithelium. This is confined to the superior part of the nasal cavity (Fig. 131). It extends medially on to the septum and laterally on to the superior turbinate. Olfactory epithelium is non-ciliated and contains the bipolar olfactory cells. The axons of these neurons combine into about 20 olfactory nerves which pass through the cribriform plate to relay in the olfactory bulb. Trauma to the cribriform plate may shear the olfactory neurons resulting in loss of smell.

Respiratory epithelium. This epithelium (Fig. 132) lines the rest of the nasal cavity. It is a pseudostratified ciliated columnar epithelium.

The respiratory epithelium which lines the nose and sinuses is the same as that lining the trachea, bronchi, and Eustachian tube. Goblet cells and mucus glands are distributed throughout the submucosa.

The entire respiratory passages function as one unit and disease processes that affect the nose and sinuses may also affect the trachea and bronchi.

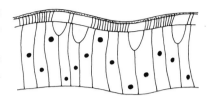

Fig. 132 The pseudostratified ciliated columnar epithelium, known as respiratory epithelium, lines the nose, sinuses, trachea, and bronchi. Muco-ciliary action keeps the respiratory tract moist and clean. Small hooks on the end of the cilia pull the viscous gel layer forwards during the effective stroke. The recovery stroke occurs mainly in the thin sol layer.

ANATOMY OF THE SINUSES

The sinuses are bony cavities within the skull. They are all formed as diverticula from the nasal cavity and are therefore lined with respiratory epithelium. The sinuses consist of the paired frontal sinuses and maxillary sinuses (antra), the ethmoid and spenoid sinuses.

The maxillary sinus (antrum)

This is the largest sinus, with a volume of 15–30 ml (Fig. 133). Rudimentary at birth, by the age of 10 years the floor of the sinus is level with the nasal floor. Mucus and debris cannot be cleared from the sinus by gravity but rely on the action of the cilia. These direct the mucus through the sinus ostium which is high up on the medial wall and opens into the middle meatus.

The roots of the premolar and first molar teeth may project into the antrum and consequently dental infection may lead to secondary sinus infection. Following dental extraction, an oro-antral fistula may arise.

Relations. The roof of the sinus forms the floor of the orbit. The infraorbital nerve runs across the roof and the bone covering it may be dehiscent. Fractures that involve the sinus roof may, therefore, lead to trapping of orbital contents or damage to the infraorbital nerve. The posterior wall of the sinus separates the antrum from the pterygo palatine fossa. The maxillary artery lies in the pterygo palatine fossa and access to this artery can be gained through the maxillary sinus.

The ethmoid sinuses

The ethmoid sinuses (Fig. 133) lie between the orbits and comprise a labyrinth of about 20 inter-communicating air cells. Nasal polyps arise from the ethmoid sinuses. Polyps may originate from each ethmoid cell which explains why they are often multiple. The ethmoid sinuses are particularly important in childhood where they account for a large proportion of the facial skeleton (see Paediatric rhinology, p. 148).

Relations. The lateral wall of the ethmoid sinus is paper thin and is consequently known as the lamina papyracea. Infection in the ethmoid sinus may spread through the lamina papyracea into the orbit.

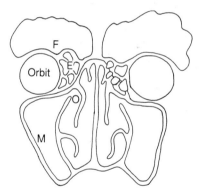

Fig. 133 Coronal section of the sinuses. The cilia direct the mucus out of the sinus ostia. Note the paper-thin bone between the ethmoid sinuses and the orbit known as the lamina papyracea. F = frontal lobes, E = ethmoids, O = sinus ostia, M = maxillary.

The roof of the ethmoid sinus forms part of the floor of the anterior cranial fossa. Damage to the cribriform plate and a CSF leak may thus occur as a complication of ethmoid surgery. Spread of infection or tumour to the anterior cranial fossa can occur through the ethmoid roof or via the ethmoidal vessels. Posteriorly the ethmoid cells lie close to the apex of the orbit and the optic nerve.

Key Points

1. Because of their relatively large size, the ethmoid sinuses are particularly important in children.

2. The ethmoid sinuses are intimately related to the orbit, the cribriform plate and the optic nerve.

Frontal sinuses

The frontal sinuses (Fig. 133) develop from a recess in the anterior part of the nose. They are rudimentary before the age of 7 years. Pneumatization into the frontal bone then occurs and its extent is extremely variable.

The frontonasal duct is a series of bony spaces that forms a passage through the anterior ethmoidal cells, down into the middle meatus. The tortuosity of this passage may explain why the frontal sinus is commonly obstructed. Mucoceles and bony osteomata may form in the sinus and obstruct the duct.

The front wall of the sinus is diploeic bone, which allows infection to track within the two layers. Osteomyelitis may follow frontal sinusitis and produce the Pott's puffy tumour.

Relations. Only a thin posterior wall separates the sinus from the frontal lobe and this may provide a route for the intracranial spread of infection. Fractures can also breach this posterior wall. The floor of the sinus is the roof of the orbit.

The sphenoid sinuses

The sphenoid sinuses lie in the body of the sphenoid bone, which is quite literally in the centre of the head. The sinuses are present at birth but are rudimentary. Pneumatization occurs from 10 years onwards and is extremely variable. The trans-sphenoidal route is a common approach for pituitary surgery and limited pneumatization makes this surgical approach more difficult. The sinus drains anteriorly into a recess just above the superior turbinate.

Relations. The lateral wall of the sinus separates it from the cavernous venous sinus whose contents include cranial nerves III–VI. The internal carotid artery also lies in the lateral wall of the cavernous sinus. During pituitary surgery, this fact is important because it is essential to stay in the midline when penetrating the anterior wall of the pituitary fossa as there is only 4 to 8 mm between the two internal carotid arteries at this point.

BLOOD SUPPLY TO THE NOSE AND SINUSES

The nose and sinuses are supplied by both the *internal* and *external carotid arteries* (Fig. 134).

The anterior and posterior ethmoid arteries

These are branches of the ophthalmic artery which in turn is a branch of the internal carotid artery and supply the superior part of the nose.

The maxillary artery

The remainder of the nose and sinuses is supplied by branches from the external carotid artery, chiefly the maxillary artery. Little's area (Kiesselbach's plexus) is an area where several vessels anastomose on the anterior septum. It is a frequent site of epistaxis.

Venous drainage

The venous drainage of the nose and sinuses is provided by the ophthalmic and facial veins and the pterygoid and pharyngeal plexuses. The venous drainage is, therefore, both intracranial and extracranial. The intracranial connection is important because infections of the face can drain via these veins to the cavernous sinus.

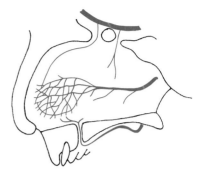

Fig. 134 Arterial blood supply of the nose. The superior part of the nose is supplied by the internal carotid and the inferior part of the nose by the external carotid.

Key Points

1. The blood supply of the nose comes from both internal and external carotid systems.

2. The veins of the nose, face and sinuses have intracranial communications which may act as routes for the spread of infection (e.g. cavernous sinus thrombosis, meningitis, frontal lobe abscess).

NERVE SUPPLY TO THE NOSE AND SINUSES

The nose and sinuses receive sensory, special sensory (smell) and autonomic nerve supplies.

Sensory

The sensory nerve supply is provided by the first and second branches of the trigeminal nerve.

Special sensory (olfactory)

The first order neurons are bipolar cells, the axons of which pass through the cribriform plate and relay at the olfactory bulb (Fig. 131). The neurons from the bulb run in the olfactory tract to the secondary olfactory centre in the frontal lobe. The cortical olfactory centre is in the dentate and semi-lunate gyri.

Autonomic

The autonomic nerve supply provides secretomotor and vasomotor control. Sympathetic fibres arise from the first five thoracic segments of the spinal cord. The fibres synapse in the superior cervical ganglion and the post ganglionic fibres run with the blood vessels to the nose and sinuses. An increase in sympathetic tone causes vasoconstriction and decreased secretion.

The parasympathetic supply to the nose is from the lacrimal nucleus with the fibres leaving the brain stem in the nervus intermedius. They relay in the pterygopalatine ganglion before entering the nasal cavity. An increase in parasympathetic tone causes swelling and increased secretion from the nasal mucosa. The pterygopalatine ganglion is sometimes called the 'hay fever ganglion'.

THE LYMPHATIC DRAINAGE OF NOSE AND SINUSES

The front of the nose and the anterior sinuses drain to the sub mandibular lymph nodes and then to the deep cervical chain. The back of the nose and the posterior sinuses drain to the retropharyngeal lymph nodes and then to the jugular nodes.

PHYSIOLOGY OF THE NOSE AND SINUSES

The nose is an organ of special sense and respiration. It is an air conditioner so that clean, warm moist air is provided to the lungs. The nose is also important in the production of sound.

Respiration

A new-born baby can only breathe nasally. If both nostrils are blocked (choanal atresia), an oral airway is required if the child is to survive. A relative dependence on nasal respiration continues into adult life, so that mouth breathing should only occur with exertion.

The nasal valve is the narrowest part of the nose. The speed of the air stream at the valve is fast and the resistance to flow is high. On inspiration, most of the air passes through the central part of the nasal cavity, up alongside the middle turbinates (Fig. 135). On expiration the airstream demonstrates turbulence (Fig. 136).

The nose as an air conditioner

The nose warms, humidifies, and cleans the inspired air.

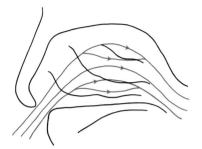

Fig. 135 Inspiratory airflow. The flow is laminar and occurs predominantly over the middle turbinate.

Heat exchange. As the air passes over the warm nasal mucosa, it extracts heat. By the time it reaches the nasopharynx, the temperature of the inspired air ranges between 30 and 34°C and is independent of the external temperature. If the external temperature falls, more blood is diverted to the nose so that the cavernous spaces within the mucosa and nasal turbinates swell. The heat output from the nose therefore increases. The reverse situation occurs when the ambient temperature rises. There is a standing temperature gradient from the front of the nose to the back, so that on expiration heat is returned to the nose.

Humidification. Even if an individual is breathing dry desert air the nose is able to humidify the air so that when it reaches the nasopharynx it is 85 per cent saturated. By the time the inspired air reaches the alveoli it is between 95 and 100 per cent saturated.

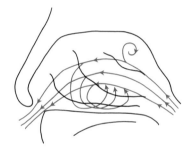

Fig. 136 Expiratory airflow. Note the turbulence.

Cleaning. The nose also cleans the inspired air. Particles over 4 μm are filtered by the nose and removed in the mucus. Smaller particles pass on to the lung and are then removed by macrophages.

The protective function of the nose

The protective function of the nose is intimately related to its air conditioning function. By cleaning and humidifying the air, the respiratory tract is protected. In addition, defence is provided by the mucus blanket, immunoglobulins, and the full range of immune cells.

The mucus blanket. This has two layers: a gel and sol phase. The top gel layer has a high viscosity and it is this layer that is moved by the ciliary beat. Each cilium is approximately 5 μm long and 0.3 μm thick and consists of microtubules arranged in a 9 × 2 pattern (Fig. 137). The recovery stroke of the cilia occur mainly in the low viscosity sol layer. Movement of the cilia is like a wheat field swaying on a windy day so that all the cilia do not move at once.

The mucus blanket moves backwards from the front of the nose to the post nasal space in about 20 minutes. In patients with Kartagener's syndrome, no mucociliary transport occurs because the cilia lack dynein

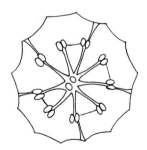

Fig. 137 Ultra structure of a cilium. Microtubules are arranged in a '9-outer–2 inner' fashion. The dynein arms contain ATPase and are deficient in Kartagener's syndrome.

arms. It is these arms which contain the adenosine triphosphate that power the cilia.

Immunological factors. The submucosa contains lymphocytes, eosinophils, mast cells, and macrophages. IgA, IgM, and IgG are also present in the nasal mucosa. In addition cells are present which secrete lysozymes, interferon, and complement factors.

Special sense: olfaction

Although poorly developed in human beings, a sense of smell adds enormously to the quality of life. The tongue taste buds allow an individual to distinguish between sweet, sour, bitter and salt. The nuances of taste are provided by the sense of smell. A loss of the sense of smell makes eating and drinking a boring experience! As well as providing pleasurable sensations, the olfactory neurons can also detect noxious substances (e.g. smoke fumes, gas, etc.).

Only volatile substances can be detected by the olfactory neurons. The molecules of these substances must be soluble in water and lipids, and about 10 to 15 molecules per mm^3 are sufficient to stimulate the olfactory neurons. The sense of smell becomes fatigued so that after a short exposure the subject may no longer be aware of the aroma.

SPEECH

In speech the nose acts as a variable resonance chamber. The palate shuts the nasopharynx when vowels are formed and opens when consonants are formed. These consonants e.g. 'm', 'n', and 'ng' can be used to test the degree of nasal patency. Nasal obstruction produces a hyponasal voice, e.g. the adenoidal child or the patient with nasal polyps.

NASAL REFLEXES

In addition to the changes in response to climatic variation, the resistance to breathing in each nostril is continually changing. At any one time one nostril tends to be blocked and the other clear. This alternating pattern is known as the nasal cycle and occurs over a two to four hour period. Other reflexes exist so that if a person sleeps on the right side, the right nostril tends to block and vice versa. Trigeminal nerve endings are present in the nose and are responsible for a diving reflex whereby the heart rate is slowed when the nose contacts cold water (try it!).

Key Points

1. Muco-ciliary transport is crucial for nasal function and occurs throughout the nose and sinuses.

2. The nose warms, humidifies and cleanses the inspired air. It is an organ of special sense (smell) and is important in speech production.

3. The nose and lungs should be considered as one functional unit.

CHAPTER ELEVEN

History and examination

HISTORY

There are six main symptoms of nasal disease: (1) obstruction, (2) discharge and post nasal drip, (3) loss of smell, (4) pain, (5) bleeding, and (6) cosmetic deformities.

Obstruction

It is often necesary to ask directly if the patient has difficulty breathing through the nose. The obstruction may be intermittent (such as the seasonal symptoms of hay fever) or permanent, and may affect one nostril or both. It may be provoked by contact with pollens, pets, dust or other antigens. Alcohol or cigarette smoke may interfere with vasomotor control and lead to obstruction. A blocked nose leads to mouth breathing and a dry throat.

The subjective awareness of nasal blockage varies considerably between individuals and does not always correlate with the assessment made on examination or with special tests.

Discharge

A discharge may be clear, yellow, green, or blood-stained. The patient may describe a clear discharge with the nose 'running like a tap' which is typical of allergic rhinitis. Rarely, a clear discharge may be due to a leakage of CSF (Fig. 138).

Most types of rhinosinusitis are accompanied by a nasal discharge. This is usually bilateral and varies from clear to purulent, depending upon the extent of infection. The discharge associated with nasal polyposis may be green due to eosinophilia or 'anaerobic' pus.

In addition to discharge from the front of the nose, patients may complain of a 'post nasal drip'. It is normal for mucus to be moved backwards into the post nasal space where it is swallowed. However, when there is an increased quantity of nasal secretions resulting from infection a foul tasting drip at the back of the nose develops.

Beware of the persistent unilateral nasal discharge. A unilateral bloodstained discharge is likely to come from a tumour. A unilateral nasal discharge in a child probably originates from a foreign body.

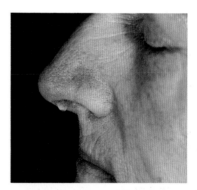

Fig. 138 Cerebrospinal fluid rhinorrhoea. Note its position just under the tip of the nose—not on the nasal floor.

Smell

Frequently a patient may complain of 'a complete loss of smell' when there is merely a reduction in smell (hyposmia). Smell and taste are closely related so that a reduction in the sense of smell may lead the patient to complain of a 'loss of taste'. Hyposmia may occur with acute rhinitis (the common cold) and is often associated with nasal polyps.

Complete loss of smell (anosmia) is unusual. It may follow a fracture of the cribriform plate where the olfactory neurons have been sheared off. It may also follow a viral infection. In both cases the loss of smell is usually permanent.

Pain

Pain from sinus disease is well localized. The exception is sphenoiditis which may present with central head pain. Characteristically, sinus pain is dull and persistent. The pain may last for several hours or days. It is aggravated by bending forwards or straining. The pressure changes associated with flying or diving may increase the severity of the pain. When the maxillary sinus is infected the pain may radiate down into the upper teeth.

Bleeding

This is dealt with in the chapter on epistaxis (p. 120). Unilateral blood-stained discharge or frank epistaxis may herald an underlying tumour and careful examination of the nasal cavities should always be undertaken in these patients.

Cosmetic deformity

It is important to consider the cosmetic aspects of the nose. The patient may be unhappy about his or her appearance, but not volunteer the information spontaneously.

EXAMINATION

The external nose

The thickness, elasticity, and general condition of the skin are important for rhinoplasty. Deviations or depressions of the nose are readily apparent. A finger passed over the dorsum of the nose allows the examiner to concentrate on defects in the nasal skeleton (Fig. 139).

Fig. 139 Gentle palpation of the dorsum of the nose detects underlying defects.

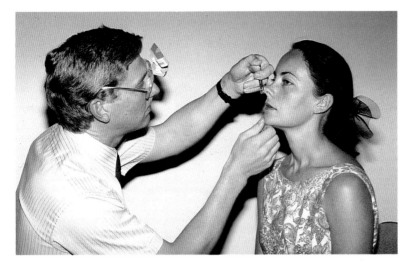

Fig. 140

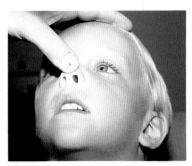

Fig. 141

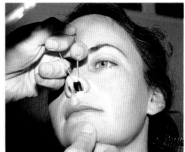

Fig. 143

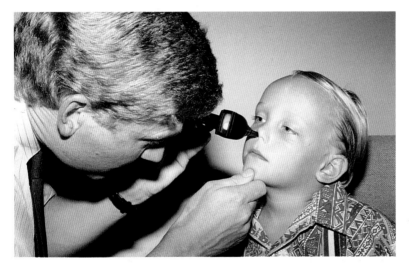

Fig. 142

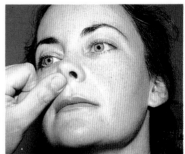

Fig. 144

Fig. 140 The examining position. Surgeon and patient are on the same level and both are comfortable. A battery operated light source which allows for normal and magnified vision is being used. This system has the advantage of being portable.

Fig. 141 A nasal speculum is not required to examine the nose of a child.

Fig. 142 An otoscope is useful for nasal examination, especially when searching for a foreign body.

Fig. 143 Gentle insertion of the Thudicum's nasal speculum.

Fig. 144 To test the patency of the nasal airway, one thumb gently occludes the nostril and the patient is asked to breathe in and out gently. The nostril being tested is not deformed.

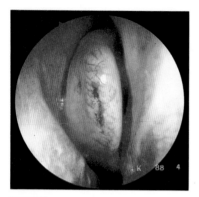

Fig. 145

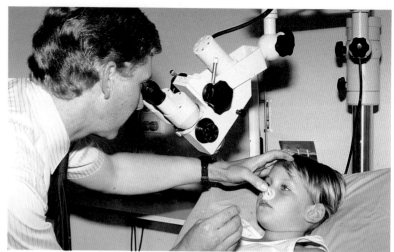

Fig. 146

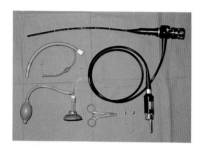

Fig. 147

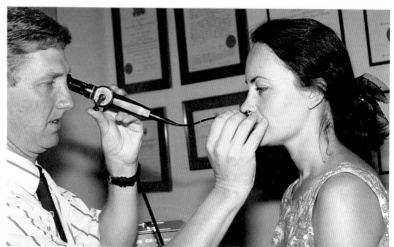

Fig. 148

Fig. 145 Nasendoscopic view of the right nasal cavity showing the turbinate (on left) and the septum (on right).

Fig. 146 The microscope being used for nasal examination.

Fig. 147 The flexible fibreoptic nasal pharyngoscope allows the nose, post nasal space, and larynx to be examined with minimal discomfort to the patient. Local anaesthetic spray is helpful. It can also help the insertion of a nasotracheal tube.

Fig. 148 The fibreoptic scope is extremely useful for assessment of the airway and can be used for intubation by 'railroading' the endotracheal tube over the scope.

The anterior nose

Examination of the nasal cavity is best done using a headlight or headmirror. A good comfortable position for the patient and the examiner is essential (Fig. 140). With young children it is usually not necessary to use a speculum. By lifting the columella a good view of the anterior nose can be obtained (Fig. 141). An auriscope with a small speculum attached is particularly useful for examining the nose (Fig. 142) when more specialist equipment is not to hand.

In adults a nasal speculum is inserted into the anterior nose (Fig. 143). The vestibule is a particularly sensitive area and this must be done gently. The blades of the speculum are held together and then gently released once the speculum is within the vestibule. The anterior septum, inferior turbinate and middle turbinate can then be examined. A normal mucous membrane is moist and pink. Septal deformities or perforations, abnormal secretions, mucosal changes, ulceration, foreign bodies, and neoplasms can be detected. Little's area at the front of the septum may reveal prominent blood vessels.

When testing the patency of a nasal airway, gently occlude one nostril with a finger (the thumb is ideal) (Fig. 144). It is important not to deform the nasal airway under test by so doing.

Table 13. ENT: history-taking and examination

Ear	Nose	Throat
History		
Earache, irritation	Obstruction	Hoarseness
Deafness	Rhinorrhoea/post nasal drip	Dysphagia
Discharge	Allergy/hay fever	Stridor
Tinnitus	Facial pain	Lump in the neck
Vertigo	Epistaxis	
	Sense of smell	Sleep disturbance
	Appearance	
Children: Speech/language		
Past History		
Barotrauma	Trauma	Cigarette smoking
Acoustic trauma	Medications	Alcohol
Head injury	*prescribed*	
Ototoxics	*non prescribed*	
Family history	Previous surgery	
Previous ear surgery		
Examination		
Pinna	Shape	Mouth
Mastoid	Septum	Larynx—refer to ENT
Ear canal	Turbinates	Neck
Tympanic membrane	Airway	
Tuning forks	?Mucopus	
Conversational test	Facial tenderness	
Nystagmus	Facial sensation	
Facial nerve	Facial swelling	

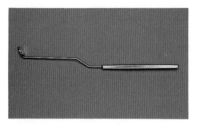

Fig. 149 The mirror used for examining the nasopharynx.

If the nose is congested, the application of ten per cent cocaine spray causes vasoconstriction and thereby improves the view.

Increasingly, rigid endoscopes are being used for clinical examination of the nose (Fig. 145). A microscope can also be invaluable (Fig. 146).

Note: A turbinate can be distinguished from a polyp by gentle palpation with a probe. A polyp is insensitive and swings back and forth on its stalk. A turbinate is a fixed sensitive structure.

The posterior nose

A flexible fibreoptic nasopharyngoscope allows the post-nasal space (nasopharynx) to be examined without any discomfort to the patient (Figs 147 and 148). Alternatively, the posterior nasal space can be examined using a small mirror inserted through the mouth (Fig. 149).

The nose and sinuses should always be examined in conjunction with the rest of the head and neck.

Table 13 summarizes the main ingredients of nasal examination.

Key Points

1. Unilateral nasal symptoms may be due to tumour or foreign body. Careful assessment is mandatory.

2. Bloodstained serosanguinous discharge may herald an underlying tumour.

3. Clear watery rhinorrhoea can be due to leakage of cerebrospinal fluid.

CHAPTER TWELVE

Investigation of nasal function and disease

....

RADIOLOGY

Introduction

The sinuses are complex air-containing cavities within the skull. The sharp contrast between air and bone allows for good radiographic images.

Plain radiographs

Plain radiographs (Figs 150 and 151) will show if the sinuses are normal or if there is gross disease (mucosal thickening, fluid levels, etc.) in any one of the sinuses. Plain films are not capable of showing precise detail or of delineating subtle pathological changes. A number of different projections are usually necessary to survey the sinuses.

Computerized tomography (CT)

CT images bony structures well, because of the high density change between air and bone. A CT scan of the sinuses is usually taken in the coronal plane and 5 mm slices are used. CT is able to demonstrate the normal structures in fine detail as well as subtle or gross pathological changes (Fig. 152).

Magnetic resonance imaging (MRI)

MRI demonstrates soft tissue abnormalities extremely well (Figs 153 and 154). Bone appears as a void. MRI has limited availability and, therefore, has not been fully evaluated in the investigation of sinus disease. It is likely to be a useful technique for delineating tumour spread.

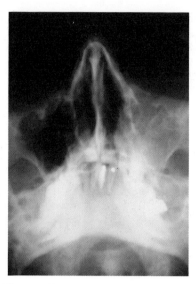

Fig. 150 Plain X-ray showing an opaque left maxillary antrum.

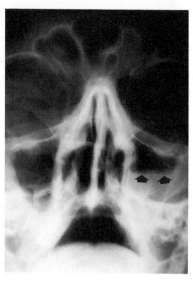

Fig. 151 Plain X-ray showing a fluid level (arrows) in the left maxillary antrum.

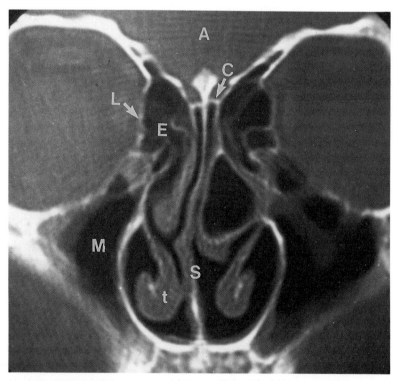

Fig. 152 Normal coronal CT of the sinuses. Note the frontal lobe (A), cribriform plate (C), lamina papyracea (L), ethmoid labyrinth (E), maxillary antrum (M), septum (S), and inferior turbinate (t).

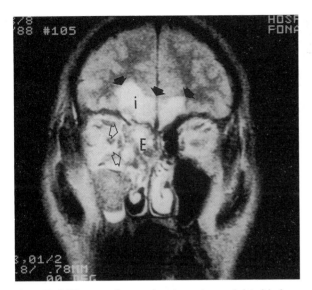

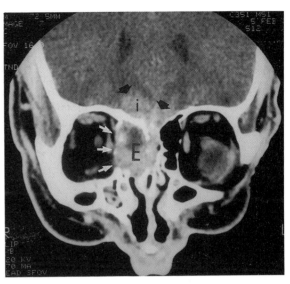

Fig. 153 MRI coronal scan showing a large right sided ethmoidal tumour (E) with intracranial (i, dark arrows) and intraorbital (clear arrows) invasion. Note that bone is not imaged. Compare with the CT in Fig. 154.

Fig. 154 CT scan. The same patient as in Fig. 153. Note how well bone is visualized but the intracranial (i, dark arrows) and intraorbital (white arrows) extensions of the tumour (E) are less clear than on MRI (Fig. 153).

Angiography

Angiography is used to demonstrate the blood supply of vascular tumours (e.g. angiofibromas). The relative importance of each feeding vessel can be demonstrated by angiography and embolization of feeding vessels can be performed to reduce tumour vascularity. Digital subtraction venous angiography allows non-invasive imaging of blood vessels.

Key Point

Plain X-rays are of limited value and are not a substitute for clinical examination of the sinuses. Mucosal thickening or fluid levels are not synonymous with infection.

RHINOMANOMETRY

This provides an objective measure of the resistance of the nasal airway. The technique measures the nasal airflow and pressure at the nostrils during respiration. These two parameters allow the nasal resistance to be calculated. There are various types of rhinomanometry but the anterior active method is the one most often chosen for routine clinical work.

Key Point

Rhinomanometry gives an objective measure of nasal airway resistance. The results do not always correlate with the patient's symptoms.

CLINICAL ENT

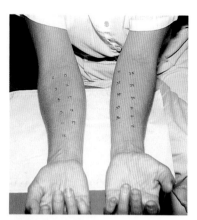

Fig. 155 Skin test for allergy. Different antigens are numbered on the forearm and the reaction to antigen is compared to a histamine control.

IMMUNOLOGICAL INVESTIGATION

Skin tests

The common allergens that cause nasal allergy are usually wind-borne. Grass and tree pollen, animal proteins (housedust mite, human dander, dog hair, and cat fur), moulds, and fungi are all implicated. The active component of these allergens can be isolated and mixed with glycerine to form a test solution.

Skin-prick tests (Fig. 155) are inexpensive, quick to perform, and a large number of allergens can be tested. There is a low risk of anaphylaxis. The rationale for skin testing is that it might identify the antigen to which the patient is allergic so that preventative measures or desensitization can be undertaken. Overall, in clinical practice, their value is quite limited and many clinicians have abandoned them.

The radio-allergo-sorbent test (RAST)

In this test a sample of the patient's serum is taken and this is mixed with a known isolated antigen. If the patient is sensitive to the allergen, the IgE in the patient's serum binds with the antigen and this is detected by adding radio labelled anti IgE. The RAST test is expensive but quantifies the allergic response and can be helpful in treatment.

THE INVESTIGATION OF MUCOCILIARY CLEARANCE

The saccharin test

This investigates the movement of mucus within the nose. A small fragment of saccharin is placed on the anterior end of the inferior turbinate. The cilia sweep the mucus and saccharin back into the pharynx, at which point the patient tastes the saccharin. The time taken for this to occur in a normal subject is about 20 minutes. If there is no mucociliary transport the saccharin stays at the front of the nose.

Brush biopsy

A brush biopsy can be taken using a standard bronchoscopy brush. This is rubbed against the turbinates and the specimen is then mixed with saline. Under a phase contrast microscope the beat of normal ciliary action can be observed. This is an important test in the investigation of mucociliary disorders but is only available in specialized centres. An early diagnosis of mucociliary disorders allows for regular physiotherapy to commence before permanent lung damage results.

IMMUNOLOGICAL TESTING

Chronic persistent sinusitis, especially in children, may result from hypogammaglobulinaemia. The levels of IgG, IgA and IgM can be measured. The importance of detecting these patients is that gamma-globulin deficiencies can be replaced.

BACTERIOLOGICAL INVESTIGATION

Routine nasal swabbing is of little diagnostic help because most of the organisms that cause sinus infection are also found in the normal nose, e.g. *Streptococcus pneumoniae*, *Haemophilus influenzae*, *Streptococcus pyogenes*.

A proof puncture of the maxillary sinus can be helpful in obtaining an accurate culture of the responsible pathogen. The proof puncture is performed by inserting a trocar and canula through the bone under the inferior turbinate (see Figs 159–161 on p. 104). The contents of the sinus are aspirated and then the sinus is washed out. Apergillosis occasionally occurs (immunosuppressed patients) and can be identified on a slide.

> **Key Point**
> A variety of investigations exist to detect allergy, immune deficiency, and mucociliary dysfunction in disorders of the nose and sinuses.

THE INVESTIGATION OF SMELL

It is possible to carry out a gross qualitative test of smell by using different scents. Some substances are detected by the trigeminal receptors and/or by the taste buds. Cinnamon is a scent that can only be detected by the olfactory neurons. It is clinically useful to know whether a smell is normal, reduced, or totally absent. Bottles of scent can provide this information.

CHAPTER THIRTEEN

Rhinosinusitis

Table 14. The causes of rhinosinusitis

(1) Allergy
(2) Vasomotor change
(3) Infection
(4) Mechanical
(5) Mucociliary
(6) Immunological
(7) Iatrogenic

Ciliated columnar epithelium extends throughout the sinuses and nasal passages which act together as one unit. Pathological changes in the nose are usually accompanied by similar findings in the sinuses, hence the term rhinosinusitis.

There is no agreed classification of rhinosinusitis. The classification is based upon aetiology (Table 14).

Allergic rhinosinusitis

Allergic rhinosinusitis is common. The antigens that affect the nose are generally windborne, e.g. grass and tree pollens, housedust mite, human dander, and dog and cat fur.

In the type 1 IgE mediated reaction, immunoglobulin E (IgE) is produced from plasma cells which in turn are regulated by T lymphocytes. IgE has a crystallizable fraction which binds to mast cells and an antigen-binding portion (Fab) which is free. When combined to an antigenic substance, the Fab portion triggers mast cell degranulation (Fig. 156). The substances released include histamine, slow reacting substance of anaphylaxis (SRS-A), leukotrienes, and prostaglandins. These substances result in the production of mucosal oedema and profuse nasal secretion.

Non-organic substances also produce an allergic response. Non-specific irritants, such as cigarette smoke and dust cause the release of vasoactive substances but the response is not IgE mediated. Sometimes the response may be due to physical factors (temperature change) alone and so is closely related to vasomotor rhinitis.

Key Point

Seasonal allergic rhinitis occurs only when pollens are in the air (hay fever). Perennial rhinitis occurs all year round and can be caused by a myriad of substances.

Vasomotor rhinosinusitis

This is a disorder of the autonomic nervous system and it is the parasympathetic side that predominates. The nasal mucosa becomes oedematous and hypersecretes in response to an environmental change. Overstimulation of the parasympathetic system leads to a blocked and running nose ('honeymoon nose'). It will be appreciated that there is considerable overlap between allergic rhinitis and vasomotor rhinitis.

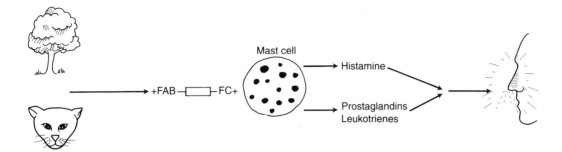

The latter is said to occur when no definite allergic response can be identified.

Infective rhinosinusitis

The best known example of infective rhinosinusitis is the common cold. A large number of viruses have been implicated (influenza, para influenza, picorna, respiratory syncytial viruses, and adenoviruses). Infection is transmitted by droplet spread. The condition resolves or a secondary bacterial infection supervenes. *Haemophilus influenzae* and *Strep pneumoniae* are the commonest offenders.

Infective rhinosinusitis is characterized by a hypersecreting and hypertrophic nasal mucosa. When bacterial infection is present the secretion becomes muco-purulent. Pus reduces the activity of cilia and this leads to stasis of secretions within the nose and sinuses.

Infection of the sinuses usually results from nasal infection although occasionally the maxillary sinus becomes infected directly from a dental abscess. The stages of sinus infection are shown in Fig. 157. Infection of the nose and sinuses can lead to serious complications, which are discussed later in the chapter.

Fig. 156 The IgE mediated hypersensitivity reaction. The antigen binds with immunoglobulin E which then attaches to the mast cell. Mast cell degranulation causes the 'hay fever' type symptoms.

Fig. 157 Stages in the development of sinus infection.

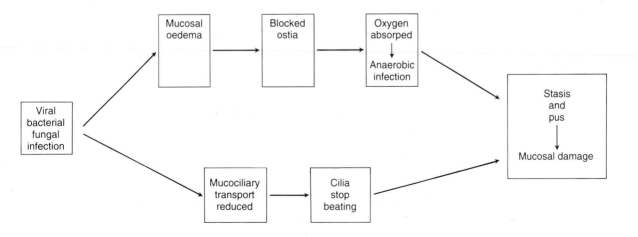

Mechanical rhinosinusitis

A deviated nasal septum may lead to or exacerbate rhinosinusitis. The obstruction produces an abnormal airflow and this allows the mucosa to dry out which inhibits the action of cilia. A deviated septum is usually accompanied by hypertrophy of the turbinates on the side away from the deviation. Obstruction of the sinus ostia may occur.

The pressure changes that occur with flying or diving (barotrauma) can lead to sinusitis if an ostium is blocked so that pressure equalization between the sinus and nasal passages cannot occur.

Mucociliary transport disorders

In some patients the cilia are abnormal (Kartagener's syndrome) or there is a defect in the mucus itself (Young's syndrome). Such disorders may present with nasal and sinus symptoms and, without prophylactic physiotherapy, bronchiectasis may develop.

Immunological disorders

The nasal passages are in frequent contact with antigens and rhinosinusitis may be the initial presentation of underlying immune deficiency.

Iatrogenic

This takes the following two forms: (a) *Rhinitis medicamentosa*. This is a condition brought about by the excessive use of vasoconstrictive sprays which can be purchased without prescription. The spray constricts the blood vessels within the nose and rapidly relieves the nasal blockage. A rebound hyperaemia occurs after two to three hours so that more spray is required to produce relief. A vicious cycle is entered with some individuals becoming dependent on the spray. (b) *Atrophic rhinitis*. This is an uncommon condition associated with excessive crusting and dryness in the nose. The nostrils become widely patent and the nose crusts and smells horribly. It may follow radical nasal surgery in which most of the nasal 'furniture' has been removed. It may occur as the end point of rhinitis medicamentosa or it may have an infective origin (*Klebsiella* is often cultured in this condition).

SYMPTOMS

The management begins by taking a history. The main symptoms of rhinosinusitis are given in the following sections.

Nasal obstruction

This may be unilateral (e.g., septal deformity), bilateral (allergy), or alternating due to the nasal cycle. In allergic and vasomotor rhinitis the obstruction may be intermittent. In allergic rhinitis there may be a clear history of the precipitating cause, for example, grass pollen, cigarette smoke, etc.

Rhinorrhoea

The discharge may come from the front of the nose or may 'drip down the back' (postnasal drip). It may be clear and watery (allergic rhinitis), or it may be purulent (infective). A unilateral seroanguinous discharge needs careful investigation in case there is an underlying tumour. A unilateral nasal discharge in a child is invariably due to a foreign body.

Facial pain

The pain associated with sinus infection is well localized and has a dull boring quality. Maxillary sinus involvement produces pain that may radiate down to the upper teeth. Ethmoidal infection typically produces pain behind the eyes. The pain of a frontal sinus infection is usually well localized over the affected sinus. Infection in the sphenoid is difficult to diagnose but produces central head pain. Sinus pain is typically aggravated by straining or lifting or by any pressure change, for example, flying or diving.

Loss of smell

A reduction in smell occurs in most types of rhinosinusitis.

Related history

Diving, flying, and any recent dental infection should be determined. Chest infections are frequently associated with rhinosinusitis.

EXAMINATION

It is unusual to have sinusitis without changes in the nose. Allergic and infective rhinosinusitis have a similar appearance. The nasal mucosa is swollen and may range in colour from pale to deep red. The turbinates are hypertrophic and there is a large amount of secretion present. The middle meatal region is a key area and pus and/or oedema is often present. In the presence of acute infection, the sinuses will be tender to percussion. Examination may also reveal septal deviation or enlarged turbinates.

Iatrogenic rhinosinusitis is characterized by a dry atrophic nasal lining. In atrophic rhinitis, the nostrils are usually widely patent even though the patient may complain of obstruction. Examine the upper teeth carefully to exclude a dental infection.

MEDICAL TREATMENT

The first principle is to treat the underlying cause.

Allergen avoidance

If there is a strong history of allergy, this can be investigated with skin tests and the patient can then be advised how best to avoid the antigens.

Desensitization

It is possible occasionally to desensitize a patient specifically, either by injection or with sublingual drops containing an extract of the antigen. The effectiveness of sublingual drops is yet to be proven and desensitizing injections carry a small risk of anaphylaxis and death.

Topical steroids and mast cell stabilizers

Topical corticosteroid drops or sprays (e.g. beclomethasone) and mast cell stabilizing sprays (e.g. sodium chromoglycate) are the most effective medical treatment for both allergic and vasomotor rhinitis. Topical steroid sprays act in a prophylactic manner and need to be taken on a daily basis. They are not associated with systemic side effects.

Antihistamines

These may help dry up the nasal secretions and more recent products (e.g. cetirizine) are non-sedating.

Antibiotics

The initial treatment of an acute infective rhinosinusitis includes appropriate antibiotics combined with a vasoconstrictor spray (Fig. 158). Ampicillin or amoxicillin are useful first line medications. Culture of pus from the nasal cavities (or from sinus wash outs) may help direct antimicrobial therapy. The rationale for a short term vasoconstrictor spray is that it can reduce swelling and open up the sinus ostia.

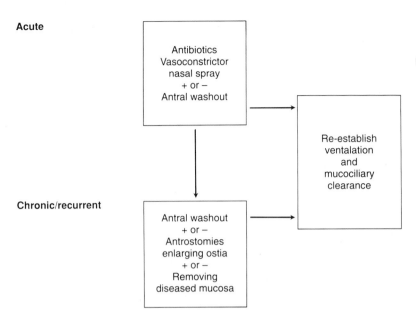

Acute

Antibiotics
Vasoconstrictor
nasal spray
+ or −
Antral washout

Re-establish
ventalation
and
mucociliary
clearance

Chronic/recurrent

Antral washout
+ or −
Antrostomies
enlarging ostia
+ or −
Removing
diseased mucosa

Fig. 158 The treatment of sinusitis.

SURGICAL TREATMENT

Sinus washouts

The maxillary sinus can be washed out as part of the treatment of acute or chronic sinusitis (Figs 159–161). Infected returns are sent for culture and sensitivity. Under general or local anaesthesia saline or water is washed through the trocar and displaces pus out through the ostium. When maxillary infection resolves it often leads to an improved drainage of the other sinuses. Antibiotic cover is advisable to prevent the occasional septicaemia that accompanies a washout.

Intranasal antrostomy

A large hole (antrostomy) can be made in the bone under the inferior turbinate or through the natural ostium in the middle meatus. This helps the drainage of the maxillary sinus.

Turbinate and septal surgery

If medical treatment of vasomotor rhinitis or allergic rhinosinusitis is not sufficient, a variety of surgical procedures are available to reduce the bulk of the swollen turbinates (cryosurgery, sub mucus diathermy, cautery, and trimming of the turbinates). If a deviated septum is present this can be corrected surgically (septoplasty). All of these methods may lead to an improved nasal airway and better access for topical sprays.

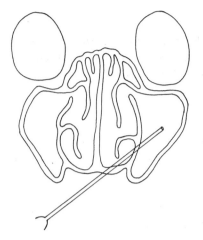

Fig. 159 Diagrammatic representation of a maxillary sinus wash out.

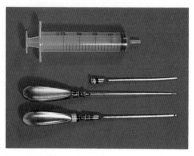

Fig. 160 Trocars, cannulae, and syringe used for sinus wash outs.

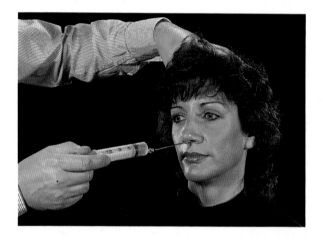

Fig. 161 The technique of antral washout which can readily be performed under local anaesthetic.

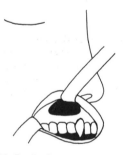

Fig. 162 Radical antrostomy (Caldwell–Luc operation). An incision is made in the gum overlying the canine fossa. The anterior wall of the sinus is removed to gain direct access to the maxillary sinus.

Radical antrostomy or Caldwell–Luc operation

The advent of endoscopic sinus surgery has all but eliminated the need for this procedure in chronic sinusitis. The operation (Fig. 162) involves removal of part of the anterior wall of the maxillary sinus to gain entry to it. Irreversibly diseased mucosa can then be removed.

Operations on the ostiomeatal complex

Fibreoptic endoscopy has made it possible to operate on the ostiomeatal complex under direct vision. The maxillary ostium, the fronto-ethmoidal duct, and the sphenoid ostium can be opened. Enlarging and clearing the sinus ostia allows ventilation of the sinuses and recovery of the normal mucociliary transport. Risks associated with this type of surgery include a CSF leak and damage to the contents of the orbit.

External ethmoidectomy

This operation (Fig. 163) is used for decompressing the orbit, as an access for pituitary surgery and for the removal of the diseased mucosal lining that occurs with nasal polyposis and chronic sinusitis.

Frontal sinus trephine

This (Fig. 164) is indicated when the frontonasal duct remains blocked despite conservative treatment. It is used as a combined procedure when intracranial pus has spread from a diseased frontal sinus.

Osteoplastic frontal flap operation

To remove osteomas, mucocoeles or diseased frontal lining, the anterior wall can be lifted off—like the lid of a biscuit tin! It also allows the posterior sinus wall to be explored after trauma.

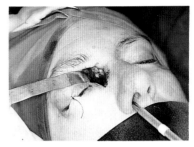

Fig. 163 Right external ethmoidectomy for decompression of the orbit. A malleable retractor allows very gentle traction to be applied to the orbital contents. The incision is made between the medial canthus and the bridge of the nose. This incision produces a nearly invisible scar.

THE COMPLICATIONS OF SINUSITIS

The complications of sinusitis can be serious and life threatening. The most common causes of intracranial sepsis are chronic sinus disease and chronic ear disease.

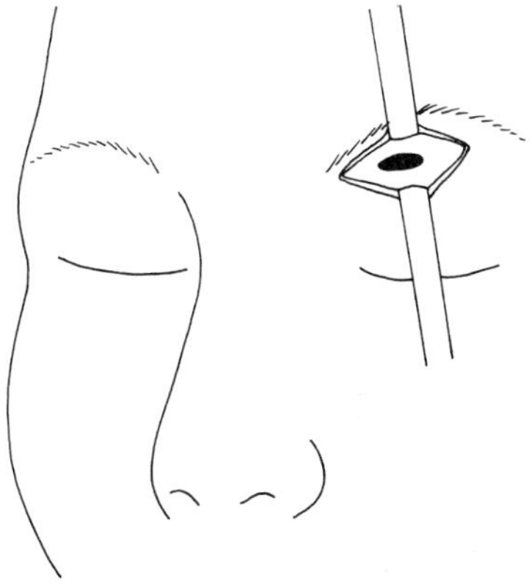

Orbital complications

These most frequently follow ethmoidal and frontal sinusitis. Pus under pressure may breach the thin lamina papyracea (Fig. 165) which forms the medial wall of the orbit. Infection can then spread to the orbital contents via veins or by breaching the periostium (Figs 166–173). Orbital cellulitis can occur within a few hours of the onset of symptoms (Table 15). Acute ethmoiditis is relatively common in childhood.

Treatment involves urgent hospital admission, intravenous antibiotics, and monitoring of visual acuity. If there is deterioration of visual acuity, the orbital contents need to be decompressed via an external ethmoidectomy (see above).

Fig. 164 Frontal sinus trephine. Through an eyebrow incision a hole is made directly into the frontal sinus. A small portex tube can be left *in situ* for temporary irrigation.

Key Point

Acute ethmoiditis can cause loss of vision. Early, intensive antibiotic therapy in hospital is necessary.

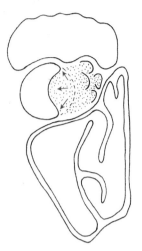

Fig. 165 Acute ethmoiditis. The potential for orbital complications is great. An abscess may cause proptosis, displacement of the orbit (downwards and outwards), and loss of vision.

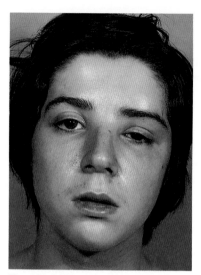

Fig. 166 Mild periorbital cellulitis early in the course of frontal sinusitis. The patient was admitted and the infection resolved on antibiotics.

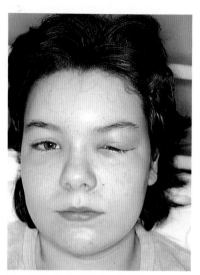

Fig. 167 Periorbital cellulitis complicating frontal sinusitis. Hospital admission is mandatory to monitor visual acuity and for intravenous antibiotics.

Osteomyelitis

Osteomyelitis of the frontal bone spreads from infection within the frontal sinuses. In young patients the anterior bony wall of the maxillary antrum may be dipleoic and osteomyelitis can follow a maxillary sinusitis.

Intracranial complications

Infection of the ethmoid or frontal sinuses may spread to the cranial cavity via a bony defect or via thrombosed vessels. *Any combination of meningitis, extradural, subdural, or intracerebral abscess is possible.* Usually a history of preceding sinusitis can be obtained from the patient or relatives.

Subdural abscess. This can be difficult to diagnose. A typical history is of a young adolescent male who has sustained frontal sinus trauma. It may not be diagnosed until the patient is unconscious because the signs are subtle initially. They include headache, pyrexia, fits, and there may be localized neurological signs. Even with a high-resolution CT scan it can be difficult to pick up such a collection of pus.

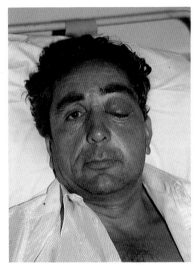

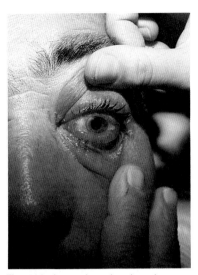

Fig. 168 An ethmoidal abscess causing swollen eyelids, proptosis, restricted eye movements, and a reduction of visual acuity. This man needed an urgent orbital decompression. See Fig. 169.

Fig. 169 Assessing visual acuity requires this manoeuvre (the same patient as in Fig. 168, see Fig. 170).

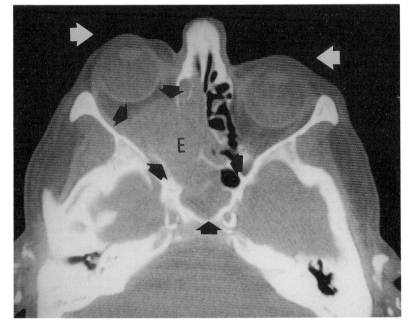

Fig. 170 Axial CT demonstrates gross proptosis (compare the positions of the white arrows) due to an ethmoidal abscess (E, dark arrows), (the same patient as in Figs 168 and 169).

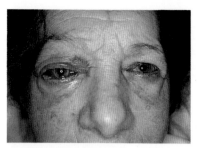

Fig. 171 Frontal sinusitis causing an intraorbital abscess. The eye was proptosed, the sclerae were haemorrhagic, and visual acuity deteriorated (see Fig. 172).

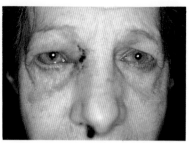

Fig. 172 The same patient as in Fig. 171 two weeks following orbital decompression.

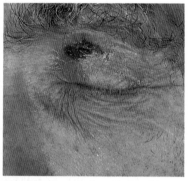

Fig. 173 Chronic frontal sinusitis caused this discharging sinus above the upper eyelid.

Intracerebral abscess. The frontal lobe is a relatively 'silent' area and it is usually this lobe that is affected. Initially the patient may not appear ill, but as the abscess forms, drowsiness and a reduced level of consciousness may occur. Fits and papilloedema may be present. Treatment requires neurosurgical consultation and involves an intensive antibiotic regime and repeated aspiration and drainage of the abscess.

Cavernous sinus thrombosis. Part of the venous drainage of the nose and sinuses is to the cavernous sinus. Thankfully, cavernous sinus thrombosis is rare, but when it does occur it follows infection in the orbit, face, nose, or sinuses. Illness may appear quickly with a sharp rise in temperature, proptosis, and pain in the eyes. There is swelling of the conjunctiva and eyelids and ophthalmoplegia develops. This condition has a high mortality.

Table 15. Presentation of the orbital complications of acute sinusitis

Swollen eyelids and proptosis
Restricted eye movements
Reduction of visual acuity

Key Points

1. Any combination of meningitis, extradural, subdural, or intracerebral abscess may follow sinus infection.

2. The onset of headache, photophobia, drowsiness, rigours, or personality change in a patient with sinus disease should always raise the possibility of intracranial suppuration.

3. A normal CT scan does not exclude intracranial suppuration, especially early in the course of the infection.

CHAPTER FOURTEEN

Nasal polyps

INTRODUCTION

Nasal polyps are pale grey sacks of mucosa that hang down into the nasal cavity (Fig. 174). They can occur at any age but are uncommon in childhood. Nasal polyps are more common in men although the sex incidence is equal in those patients who also have asthma.

Most nasal polyps are bilateral. Beware of the unilateral polyp; it always needs to be biopsied as it may be an inverting papilloma or a carcinoma. Nasal polyposis is a chronic condition with a high incidence of recurrence, whichever treatment regime is pursued. Although the disease process mainly affects the ethmoids, polypoid change can occur elsewhere in the nose and other sinuses.

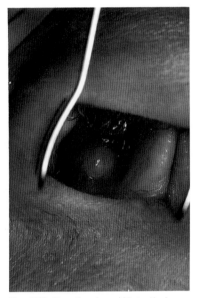

Fig. 174 Nasal polyps. Note their typical pale gelatinous appearance.

Aetiology

The aetiology of nasal polyps is not fully understood. There is no evidence that infection causes nasal polyps although infective sinusitis may result from polyps blocking the sinus ostia. An allergic aetiology has been proposed because 90 per cent of polyps contain an increased number of eosinophils. Eosinophils, however, are not specific for allergy and positive skin tests are no more common in patients with nasal polyps.

A consistent finding is that the fluid inside a polyp contains a high level of histamine (100 to 1000 times that of serum). This implies that mast cell degranulation occurs. The products of mast cell release may not be cleared from the ethmoid mucosa as rapidly as from other tissues because of the relatively poor blood supply. In addition to allergy, several other factors can cause mast cell degranulation, e.g. temperature change, various drugs, and complement factors.

Twenty five per cent of patients with nasal polyps develop asthma and about the same proportion of late onset asthmatics develop nasal polyps. Treating the nasal polyps reduces the nasal resistance and usually improves the chest symptoms. In 8 per cent of patients with nasal polyps there is a triad consisting of nasal polyps, asthma, and aspirin sensitivity.

In children, polyps should suggest a meningocoele (p. 148) or cystic fibrosis.

Clinical features

The most common symptoms are nasal obstruction, rhinorrhoea and post nasal drip. The discharge may be clear or green depending on the degree of eosinophilia or infection. Sneezing and loss of smell are also frequent but pain is unusual unless the sinuses are secondarily infected. Bleeding or a serosanguinous discharge is unusual and a carcinoma should be suspected.

On examination, the voice is hyponasal, as if the person has a permanent cold. Polyps are bilateral and appear as pale, glistening grey sacks hanging down into the nose. It is important to distinguish between a polyp and a turbinate bone: the former is insensitive and mobile when palpated with a probe. Extensive polyps may protrude from the nose and in extreme cases the nasal bones may be splayed apart producing a frog face. The signs and symptoms of nasal polyps are variable because polyps can undergo spontaneous regression without treatment.

Investigations

Positive skin tests are no more common in people with polyps, but a concomitant allergy to house dust or pollen may exist and advice can be given accordingly.

Sinus radiographs are of little value. They may show opaque ethmoid or maxillary sinuses—this does not necessarily indicate infection.

In a child an encephalocoele should be excluded (by CT scanning) and the sweat test should be performed to identify cystic fibrosis.

MEDICAL TREATMENT

Topical steroids

At least 50 per cent of people with nasal polyps have a good response to topical steroids (beclamethasone). At the moment it is not possible to identify the responders before treatment. If the polyps are causing minimal symptoms it is reasonable to use topical steroids as a first line of management. A three month course with review is appropriate.

The most effective method of applying beclamethasone is in the form of drops administered when the patient is in the 'Mecca' position (Fig. 175). In this position the steroid reaches the ethmoid cells. After an initial treatment the patient can change to an aqueous spray of beclamethasone. The spray needs to be used each day on a prolonged daily basis (leaving the spray by the toothbrush is a good reminder). There are few unwanted effects associated with topical steroid sprays because only a small fraction of the dose is absorbed systemically.

Fig. 175 The 'Mecca' position is the best position for getting nasal drops to reach the ethmoid cells.

Systemic steroids

Providing there are no contraindications, systemic steroids have a role in the treatment of severe polyposis and are relatively safe if used in a short reducing dose.

SURGICAL TREATMENT

Nasal polypectomy

The majority of patients with polyps require a polypectomy (Fig. 176) which clears the bulk of the disease. A nasal polypectomy can be carried out under a local or general anaesthetic. Polypectomy removes the polyps as near as possible to their 'roots' in the ethmoid cells.

Ethmoidectomy

This procedure is aimed at clearing the ethmoid cells from which the polyps arise. Ethmoidectomy (p. 105) is thought to reduce the recurrence rate of polyps but there is no hard evidence to substantiate this.

Postoperative nasal sprays

Topical steroids given postoperatively reduce the rate of recurrence. The problem is how long to keep the patient on the steroid spray. This will depend on the severity of the polyps, the age of the patient and how often they have recurred. A minimal time is probably three months.

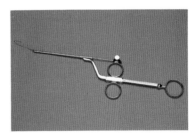

Fig. 176 A nasal snare is used to remove polyps.

ANTROCHOANAL POLYPS

These are unilateral polyps that arise from the lining of a maxillary sinus. The polyp is single and arises in the maxillary antrum (Fig. 177). It protrudes through the maxillary ostium and projects backwards to the post nasal space (nasopharynx). The polyp has a dumbell shape because of constriction at the sinus ostium (Fig. 178). The aetiology is unknown and the polyp is benign.

The most common symptom is nasal blockage which is unilateral unless the polyp is large enough to block both sides of the post nasal space.

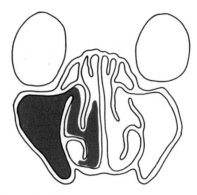

Fig. 177 The antrochoanal polyp is a single unilateral polyp arising from the maxillary sinus. It protrudes through the sinus ostium and hangs down into the postnasal space (nasopharynx).

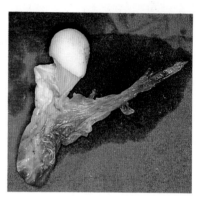

Fig. 178 This antrochoanal polyp has been avulsed. Above lies the round, smooth portion that went backwards into the nasopharynx and below it are the 'roots' which were pulled from the maxillary sinus.

Treatment

The treatment is surgical removal.

Key Points

1. Be suspicious of the unilateral nasal polyp—it may be malignant. Always send for histology.

2. Nasal polyps are like weeds—they tend to recur.

3. Polyps in children: consider meningocoele or cystic fibrosis.

CHAPTER FIFTEEN

Fractures of the nose and mid-face

NASAL FRACTURES

This injury is common in contact sports and fights. If the blow is from the front, the nasal bones may be depressed. Often the blow comes from the side so that the nasal bones are deviated. In children, the possibility of non-accidental injury may need to be considered.

A nasal fracture may be simple (skin and mucosa intact) or compound (exposed bone or cartilage).

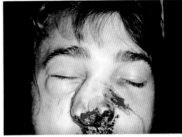

Fig. 179 This compound nasal fracture needs debriding under a general anaesthetic. This patient also had an orbital fracture.

Symptoms

The patient complains of nasal deformity, obstruction, and bleeding.

Examination

On examination, the deformity is usually obvious (Figs 179 and 180). Careful palpation will reveal whether or not the nasal bones are depressed and which, if any, are deviated. It is important to examine the nasal cavity to see if a septal haematoma is present. Radiography is not required for an isolated nasal fracture.

Treatment

A nasal fracture can be reduced under a local or general anaesthetic. This can be done immediately, i.e. before the onset of swelling or after seven days when the swelling has subsided.

Like the reduction of a Colles fracture, the deformity is at first accentuated so that the bony fragments are disimpacted. Swinging the nose back towards the midline then reduces the fracture. If a nasal bone is depressed it can be lifted using an elevator (Fig. 181). If the nose is unstable following reduction, a plaster is applied for a week.

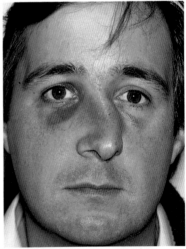

Fig. 180 A sideways blow from the right has fractured the nose and produced this deformity. Always ensure the zygoma has not been fractured.

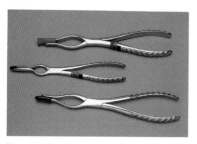

Fig. 181 Various forceps used to reduce a nasal fracture.

COMPLICATIONS OF NASAL FRACTURE

Septal haematoma

A septal haematoma (Fig. 182) occurs when there has been a shearing injury stripping the perichondrium from the underlying cartilage. The blood supply to the cartilage comes from the perichondrium and the cartilage may necrose if the haematoma is not drained. If an abscess forms in the haematoma (Fig. 183), the destruction of cartilage is accelerated and intracranial sepsis may follow. Cartilage necrosis is disastrous because subsequent fibrosis causes collapse of the nasal bridge ('saddle nose') and gross nasal obstruction.

A septal haematoma can be diagnosed by examination. The septum, which is usually thin and firm, becomes like a bluish sausage. Palpation with a probe will confirm the swelling is fluctuant.

A haematoma needs urgent drainage and firm nasal packing so that the perichondrium adheres to the cartilage.

Septal deviation

It is hard to correct deformed cartilage definitively with simple manipulation as its 'springiness' tends to make it dislodge (Fig. 184). An elective septoplasty may be required.

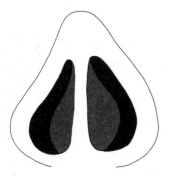

Fig. 182 A septal haematoma follows trauma. A boggy, fluctuant swelling is seen. A septal haematoma needs to be drained and the nose packed. Antibiotic cover is advisable.

Haemorrhage

This may be torrential and difficult to control. The management of epistaxis is discussed on p. 122.

Nasal deformity

Nasal fractures are frequently neglected (Fig. 185). Failure to correct adequately the initial deformity may necessitate difficult secondary procedures (septorhinoplasty) later.

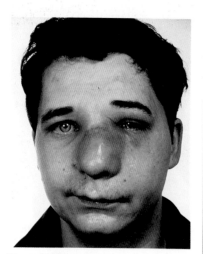

Fig. 183 This patient developed a septal haematoma followed by a septal abscess.

Key Points

1. In a patient with a nasal injury, always examine inside the nose to exclude a septal haematoma. If present, a septal haematoma needs urgent treatment.

2. Nasal fractures may be reduced immediately or after 7–10 days.

3. Adequate initial treatment of fractures will remove the need for difficult reconstructive procedures later.

4. A concomitant fracture of the facial skeleton, e.g. of the zygoma, should always be excluded.

5. X-rays of the nasal bones are not necessary in the management of simple nasal fractures.

SEPTAL DEVIATIONS

Aetiology

Septal deviation may result from genetic factors, birth trauma, or injury later in life. The deviation may be either in the cartilaginous or bony components or both.

Clinical features

The usual symptom is of unilateral nasal obstruction. If the nose is deviated (Fig. 185), as is commonly the case, it may be cosmetically unacceptable. A patient with a deviated septum is predisposed to sinusitis and the altered airflow may also lead to a dry mucosa which subsequently bleeds.

Patients with septal deformity vary greatly in terms of their symptoms. Some with a grossly deviated septum do not have any problems. Others feel that the nose is blocked when there is only minimal septal deviation present.

As assessment of the degree of nasal obstruction can be made clinically (p. 89). An objective assessment can be made using a rhinomanometer.

Treatment

A slight septal deviation will not need treatment—remember that few septums are strictly in the midline!

If there is cosmetic deformity or nasal obstruction, surgical treatment is recommended. There are two main operations: *septoplasty* and the *submucus resection* (*SMR*) of the septum. In septoplasty, the cartilaginous excision is more conservative.

The complications of septal surgery include septal perforation, collapse of the nasal dorsum, and adhesions between the septum and the lateral wall of the nose.

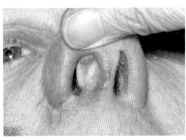

Fig. 184 The caudal end of the septum is displaced into this man's right nostril. Gentle elevation of the columella reveals the deformity.

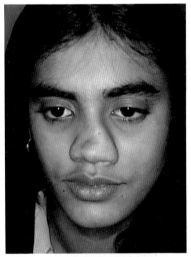

Fig. 185 This nasal deformity resulted from an untreated nasal fracture. Primary reduction is strongly recommended to prevent the need for difficult secondary reconstructions such as shown here.

SEPTAL PERFORATIONS

Aetiology

The causes of septal perforations are shown in Table 16. The most common cause is post-surgical trauma (Fig. 186). This is thought to occur when a mucosal flap is ripped on both sides and is more likely following submucus resection. Syphilis produces a perforation in the bony septum. Aggressive cautery to the septum can lead to a perforation and for this reason it is wise to cauterize just one side of Little's area at a time.

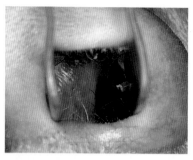

Fig. 186 Septal perforation. The most likely cause of this is previous nasal surgery.

Table 16. Causes of septal perforation

1. Trauma
   Post surgery, (submucus resection, septoplasty)
   Nasal cautery
   Pressure from nasal balloons
   Nose picking

2. Chronic infections
   Syphilis, tuberculosis

3. Malignant tumours
   Melanoma, carcinoma, lymphoma

4. Chemical trauma
   Cocaine addiction, industrial (e.g. chromate workers)

Clinical features

A small perforation may present with whistling as the patient breathes in and out, but this is lost as the perforation enlarges. A septal perforation may present with nasal obstruction, because crusting occurs around the edges of the perforation. It may also present with bleeding because crusts become detached leaving vessels exposed. A severe septal perforation leads to collapse and fibrosis, so that the nasal dorsum is depressed (saddle nose).

Treatment

No treatment is required if the perforation is asymptomatic. Saline douches and the application of vaseline to the edges is helpful in combating excessive crusting. Granulation tissue around a perforation should always be biopsied.

The surgical closure of a septal perforation is difficult. Many operations are described which rely on transposing mucosal flaps. An alternative is to insert a specially made silastic button into the hole, but this is not tolerated by all patients.

Key Points

1. Malignant disease and granulomas can present with a perforation and biopsy of any abnormal tissue should be routine.

2. Aggressive cautery to the nasal septum can result in a septal perforation.

FRACTURES OF THE ZYGOMA AND THE BONY ORBIT

Combined fractures of these structures are common (Fig. 187). The mechanism of fracture is usually a blunt violent injury to the lateral part of the face.

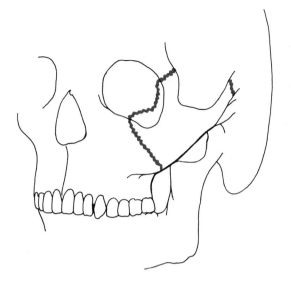

Fig. 187 The tripod fracture, so named because the fracture line runs through the zygomatic arch, the orbital rim and the zygomatic process of the frontal bone. Check integrity of orbital rim, infra-orbital nerve, and eye movements.

Examination

Inspection may reveal the orbit at a slightly lower level than its fellow at the opposite side. Palpate the orbital rim to see if the fracture has produced a step. A depressed maxilla may not be obvious because of soft tissue swelling. There may also be parasthesiae or numbness in the distribution of the infraorbital nerve. Carefully check the patient's jaw movements. Examination of the eye movements, particularly upward gaze, can detect orbital entrapment.

Treatment

A non-displaced fracture without any complications does not require treatment. Orbital entrapment, a displaced fracture, or infra-orbital nerve damage require reduction.

An uncomplicated fracture of the zygoma simply needs elevation. An unstable fracture may need open reduction and wiring.

ISOLATED 'BLOW OUT' FRACTURE OF THE ORBIT

This is often caused by a ball hitting the orbit directly. The squash ball is notorious because it is just the right size to fit in the bony orbit. The floor of the orbit is a weak spot which fractures under pressure and may lead to entrapment of orbital contents. The orbital fat, inferior rectus and inferior oblique muscles may be trapped.

Such entrapment produces enophthalmus, double vision and limitation of eye movement.

A radiograph shows a typical 'tear drop' sign which is due to soft tissue trapped in the fracture.

Entrapment requires surgical exploration. The orbital contents are freed and a silastic prosthesis can be placed on the orbital floor to hold the contents in place.

FRACTURES OF THE MIDDLE THIRD OF THE FACE

Fractures of the middle third of the face can compromise the airway. These are most commonly transverse and result from direct trauma to the face. Three typical fracture lines were worked out by Le Fort who dropped cement bags from the ramparts of a castle on to cadaver heads (Fig. 188).

Le Fort I

Here, the upper alveolus is detached. The patient has a normal occlusion and blood may be present in the antrum.

Le Fort II

In this the nasal pyramid and the upper jaw is detached. The fracture passes through the nasal bones, but below the cheek bones (zygoma). The ethmoid sinuses, orbital contents, and lacrimal apparatus may be involved.

Le Fort III

In these fractures, the facial skeleton becomes separated from the base of the skull. There is a massive depression of the middle third of the face, producing the so-called 'dish face'.

Fig. 188 Le Fort fractures. Le Fort I: The upper alvoelus is detached; Le Le Fort II: The entire upper jaw is detached; Le Fort III: The facial skeleton is separated from the skull base and may obstruct the airway.

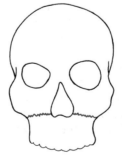  

Middle third fractures result from high speed injuries and this is most likely to occur with motor vehicle accidents. Careful history and examination are important. The middle third of the face is inspected and careful palpation around the facial structures is made. It is essential to check the full range of movements of the lower jaw (mandible). Any mobility of the upper jaw (maxilla) is clearly abnormal. It is important to examine the eye, and eye movements.

Management

These patients often have multiple injuries and need urgent resuscitation. The first priority is the airway because the tongue or a mobile maxilla may obstruct it. A useful immediate measure is to pass a Foley catheter into the nasopharynx via the nose. By inflating the balloon and applying traction, the maxilla can be pulled forward. A tracheostomy may be required if endotracheal intubation is unsuccessful.

Reduction and fixation of the facial fractures is undertaken when the patient's condition permits.

Severe fractures require the combined efforts of neurosurgeons, oral–maxillofacial surgeons, and otolaryngologists.

FRACTURES OF THE MANDIBLE

With severe fractures of the mandible the priority is to maintain the airway and stop haemorrhage. Considerable force is required to fracture the mandible so it is important not to overlook intracranial or other injury.

Examination

Look and feel both inside and outside the mouth for deformity and malocclusion, and test for labial sensation (inferior dental nerve). An orthopantomogram (OPG) is the most helpful radiograph.

Treatment

Undisplaced fractures with no malocclusion do not require operative treatment. This is usually the case with fractures of the condyle, ascending ramus and angle of the mandible. Analgesia and antibiotics are sufficient.

A *simple fracture* that reduces easily can be held in place by inter-maxillary fixation. A common method of achieving this is to wire the teeth of the maxilla to those of the mandible (eyelet wiring). An *irreducible fracture* or one that is unstable requires an open reduction and fixation. After reduction the fragments are held in alignment with either wires or by compression plating.

> **Key Points**
>
> **1.** In orbital trauma, check eye movements, palpate the bony orbital rim and record visual acuity.
>
> **2.** In patients with facial injury, always check the full range of jaw movements and determine whether or not the upper jaw is mobile. Fractures of the cheek bone (zygoma) are often overlooked.

CHAPTER SIXTEEN

Epistaxis

....

INTRODUCTION

A nose bleed (epistaxis) should never be underestimated—the blood loss can be life-threatening.

In young people the bleeding site is generally on the anterior part of the nasal septum, just behind the columella (Little's area) (Fig. 189). With increasing age the site of bleeding moves further back in the nose. Posterior bleeds are usually more severe and can be difficult to control.

Epistaxis may be due to local or general causes (Table 17). Patients with hypertension do not bleed more often but when an epistaxis occurs, it is more severe. Hypertensives, therefore account for an increased proportion of hospital admissions for epistaxis.

Table 17. The causes of epistaxis

Local causes

1. Trauma
   Nose picking
   Fractures—nasal bones, skull base, sinuses
   Following nasal surgery

2. Idiopathic
   Posterior bleeds (>30 yrs)—vessel 'degeneration'

3. Neoplastic
   Bleeding polyp septum
   Angiofibroma
   Carcinoma—nose, sinuses, post-nasal space
   Lymphoma
   Wegener's granuloma

General causes

1. Medication
   Anticoagulants

2. Haematological disease
   Familial hereditary telangectasia (Osler's disease)
   Bleeding diatheses

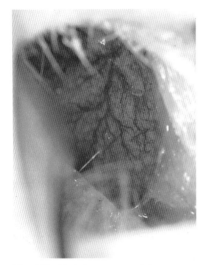

Fig. 189 Little's area—a rich anastomosis of vessels on the anterior part of the septum—a common site for epistaxis.

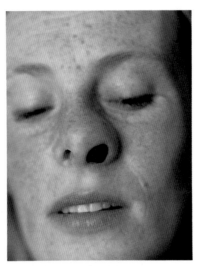

Fig. 190 Bleeding polyp of the septum (left side). This polyp enlarged during pregnancy.

LOCAL CAUSES OF BLEEDING WITHIN THE NOSE

Nose picking

The anterior part of the septum (Little's area) (Fig. 189) is an area where the mucosa may dry, particularly if there is a septal deviation. If crusting occurs the vessels may bleed when the crust is picked away. This type of bleeding is particularly common in childhood. In children there is a small vein anterior to Little's area which is often responsible for the epistaxis. In adults anterior bleeding is from an artery in Little's area.

Ageing

The posterior bleeds of advancing age are thought to be due to degeneration of blood vessels. The muscular tunica media is replaced with fibrous tissue and calcification occurs in the larger arteries.

Bleeding polyp of the septum

A bleeding polyp of the septum is occasionally seen on the anterior third of the septum (Fig. 190). This is a red angiomatous neoplasm which is thought to be caused by mechanical irritation and is treated by excision.

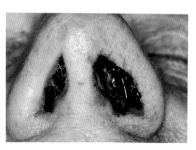

Fig. 191 Epistaxis. The patient presented with left sided nose bleeds. Examination confirmed a malignant melanoma.

Neoplasms

A carcinoma of the nose or sinuses often presents with a serosanguinous discharge rather than fresh blood (Fig. 191). One of the presenting symptoms of a nasopharyngeal carcinoma is epistaxis. An angiofibroma is an uncommon tumour that occurs in adolescent males and arises in the posterior part of the nose. It presents with profuse and recurrent epistaxes.

Trauma

Facial trauma can give rise to persistent epistaxis, particularly where fragments of the fracture overlap. When the fracture is reduced the epistaxis tends to settle.

SYSTEMIC CAUSES OF EPISTAXIS

Familial hereditary telangiectasia (Osler's disease)

This is a systemic disease (Fig. 192) in which there is a deficiency in the endothelial layer of blood vessels. Telangiectases are present on the face, lips, buccal mucosa, and tongue. Bleeding can occur at several points throughout the aero-digestive tract. The epistaxes can be difficult to control. Skin grafting of the septum or oestrogen therapy may be needed.

Oral anticoagulants and blood dyscrasias

Control of the epistaxis in these patients requires collaboration with the patient's haematologist.

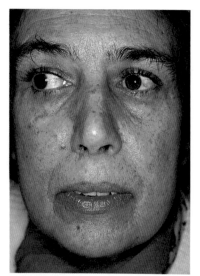

Fig. 192 Hereditary haemorrhagic telangiectasia (Osler's disease). The patient experienced profuse nose bleeds. Note the facial telangiectasia. The mainstay of treatment is oestrogen therapy.

THE MANAGEMENT OF EPISTAXIS

History

The patient may need immediate resuscitation but generally it is possible to take a history to elicit local or systemic causes for the epistaxis and to assess the amount of blood loss. Always enquire about the patient's medication (anticoagulants, aspirin ingestion, etc.).

General measures

It is important to have a positive attitude and to appear calm and in control. It may be a life-threatening condition which the patient and possibly the nurse have never encountered, but (hopefully) *you have*.

Practical measures are the same as for any patient following blood loss. Depending on the severity of the bleed, vital signs must be monitored. Blood should be taken for full blood count, cross matching, and coagulation studies as indicated. Intravenous fluid replacement may be needed.

Examination

The doctor should wear gloves, goggles and an apron to protect from contact with blood. Any blood clots in the nose need to be removed (ask the patient to blow the nose, or use a sucker). The nose should be anaesthetized with lignocaine or cocaine spray (the latter has the advantage of being a powerful vasoconstrictor).

Electrical or chemical cautery

Bleeding vessels can be cauterised with silver nitrate or electrocautery after local anaesthesia. If silver nitrate (Fig. 193) is used it is imperative to cauterize around the vessel and leave the stick in contact with the mucosa for at least ten seconds. Care is needed to prevent any nitric acid running out on to the skin which can cause a chemical burn.

Electrocautery (Fig. 194) has its dangers. It is easy to burn the skin of the nose if one is not extremely careful. Always test the cautery *before* inserting it in the patient's nose. *Do not introduce a red hot cautery into the nose—this terrifies the patient and makes a burn more likely.* Switch on the cautery only when the tip is actually on the bleeder and switch it off before removing it from the nostril.

If there is no obvious bleeding site in the front of the nose, the posterior nose can be examined with the nasopharyngoscope or the microscope. Cauterizing the more distant reaches of the nose is difficult and an operating microscope with a 300 mm objective lens is helpful.

Fig. 193 Silver nitrate cautery. On the wet mucous membrane, the silver nitrate is converted to nitric acid (on right). Excessive nitric acid could burn the skin and should be dry mopped following cauterization of the vessel.

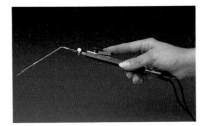

Fig. 194 Electrocautery. To prevent burns to the skin, it should only be turned on inside the nose when it is in contact with the bleeding point. Before removing the cautery, switch off the current.

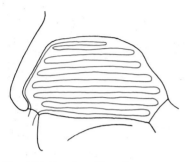

Fig. 195 Nasal packing. Remove blood clot; anaesthetize the nose with cocaine. The pack is inserted by building it up in layers from the floor of the nose upwards. A small pack just poking into the nostril is useless. Roughly one metre of packing can be inserted into the average nostril.

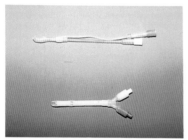

Fig. 196 Two examples of nasal balloons. The Brighton balloon (above) and the 'semi-rigid' tube balloon (below). The latter is most useful for home visits and other situations where facilities are limited.

Packs and balloons

If a bleeding point cannot be identified and controlled a nasal pack may be necessary. Bismuth iodoform paraffin paste (BIPP) is commonly used to pack the anterior part of the nose (Fig. 195). A ribbon gauze impregnated with adrenaline and cocaine is also suitable.

A balloon is an alternative to a pack and various types are available (Fig. 196). Once the nose is packed or a balloon inserted, hospital admission is essential in case the pack dislodges and obstructs the airway. A nasal pack may cause a reflex hypoxaemia, which can be important in patients with respiratory insufficiency. Depending on the severity of the bleed it is usual to leave a pack or balloon in place for 24 to 48 hours. Antibiotic cover is advisable.

Bedrest and sedation are an important part of treatment for severe epistaxis. An elevated blood pressure usually settles with bedrest. If the diastolic pressure remains high the advice of a physician is essential. A summary of the management of epistaxis is given in Table 18.

Table 18. The treatment of epistaxis

General measures
Monitor vital signs
Blood for FBC, cross match, coagulation studies
Intravenous infusion

Local measures
Topical anaesthetic spray
Cautery (electric or chemical)
Packing: BIPP, balloons, tampons
Vessel ligation (external carotid, maxillary, or ethmoidal arteries)
Septal surgery

RECALCITRANT EPISTAXIS

A persistent or recurrent epistaxis presents a challenge (Fig. 197). The exact method of dealing with the problem will depend upon the individual patient but a number of options are available.

Firm packing of the posterior nose

This is combined with packing the anterior nose and requires a general anaesthetic.

Septoplasty

This helps to gain access to the bleeding site. The operation itself can sometimes prevent recurrent epistaxis because the vessels become fibrosed after mucosal flaps have been raised and replaced.

Ligation of feeding vessels

The external carotid artery can be ligated in the neck. Alternatively, the maxillary artery can be clipped as it courses behind the maxillary antrum. Bleeding predominantly from the superior part of the nose requires ligation of the ethmoidal vessels on the medial wall of the orbit (see external ethmoidectomy, p. 105). Vessels can also be selectively embolized using invasive radiological techniques.

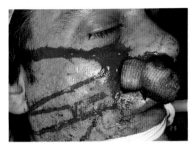

Fig. 197 This patient bled profusely despite tight packing. His maxillary artery was subsequently ligated.

Key Points

1. Epistaxis can cause life threatening blood loss. Systemic and local measures are necessary in the management of these patients.

2. It is safer to admit patients to hospital following insertion of a nasal pack. A pack can dislodge and be aspirated.

3. Do not cauterize both sides of the septum at one sitting.

CHAPTER SEVENTEEN

Facial pain

....

The management of facial pain is a challenge, both diagnostically and therapeutically. The occurrence of referred pain and the complex nerve supply to the head and neck mean that a thorough history and examination are paramount for the accurate diagnosis of facial pain. A working classification is helpful (Table 19).

NEURALGIA

A neuralgia is a pain in the area of distribution of a sensory nerve.

Primary neuralgias

A primary neuralgia is characterized by a sharp severe pain, rather like a stab with a knife that lasts for a few seconds. There may be trigger zones which stimulate the pain, but there are *no objective signs of impaired nerve function*.

Many explanations have been proposed to explain primary neuralgia. It is thought to be similar to epilepsy in that a sudden irregular discharge of nerve impulses occurs.

The quality of the pain and its distribution are the hallmarks of the diagnosis. The pain is usually unilateral. Paroxysms of pain occur in cycles and last for weeks or even months. The pain is so severe that grimaces of the face occur (gargoyles may have been modelled on patients who had trigeminal neuralgia!)

Trigeminal neuralgia. In trigeminal neuralgia the pain is restricted to the area supplied by the three branches of the nerve. The maxillary and mandibular branches are most commonly affected. There is often a trigger area. For example, the patient describes touching or brushing the teeth as the trigger which sets off the pain. Many patients have had unnecessary dental extractions and sinus washouts before the diagnosis is made.

Glossopharyngeal neuralgia. This is much rarer than trigeminal neuralgia. It follows the distribution of the glossopharyngeal nerve which carries sensation from the pharynx and tonsils, posterior third of the tongue, Eustachian tube, middle ear, and external auditory meatus. It is in this distribution that pain is felt and paroxysms are triggered by swallowing, chewing, or yawning. The trigger zones tend to be the tonsils and pharyngeal wall.

Thorough investigation is required to exclude a secondary neuralgia before a diagnosis of primary neuralgia is made. This may involve radiology of the skull base to exclude nerve compression.

Treatment. Antiepileptic drugs, e.g. carbamazepine, are usually effective at controlling primary neuralgias. The carbamazepine is started at a low dosage and gradually increased. Surgical treatments range from local nerve avulsion to ablation of the sensory ganglion. The help of a pain clinic can be invaluable.

Secondary neuralgias

This may present like a primary neuralgia initially, but then objective signs of nerve pathology occur. A common finding would be parasthesiae and numbness in the distribution of the nerve involved.

Pain in the distribution of a cranial or spinal nerve accompanied by signs of nerve impairment needs complete investigation to find the underlying pathology. A thorough examination can be supplemented by CT or MRI scans.

REFERRED PAIN

Stimulation of a sensory nerve usually produces pain in the peripheral terminals of the nerve. The pain is referred when it is felt in distant nerve terminals that share the same central connection. There are many examples of referred pain in the head and neck, e.g. irritation of cranial nerve IX (post-tonsillectomy) causes referred otalgia. A laryngeal carcinoma can also present with referred otalgia. Both cranial nerves IX and X have branches that supply the ear as well as the pharynx and larynx, respectively.

SINUS PAIN

With the exception of sphenoiditis, sinus pain is well localized. It has a dull character and lasts for several hours. It is made worse by pressure change, bending or straining and may be associated with mucopurulent rhinorrhoea. Local tenderness can be elicited by percussing the sinus.

PAIN OF DENTAL ORIGIN

This can be difficult to diagnose as occasionally the only finding is localized tenderness on chewing. The pain usually presents locally but it may also be referred to the ear. Radiological examination, particularly the orphopantamogram, is useful to detect a root abscess. Caries, periapical disease, peridontal abscesses, osteomyeilitis, fractures, and neoplasms in the maxilla and mandible are all causes of facial pain.

TEMPOROMANDIBULAR JOINT PROBLEMS

This usually occurs when there is a malocclusion which puts strain on the joint capsule. Often no changes can be demonstrated on radiographs. A temporomandibular joint disorder may present with otalgia and occasionally other otological symptoms. The pain typically radiates from the joint up towards the temple. It may be related to chewing but is not always so. Palpation may reveal malocclusion, joint tenderness or subluxation.

CERVICAL SPINE DISORDERS

These are very common and range from a mild spondylosis to severe arthritic change narrowing the foramina in the cervical spine. Even with moderate changes in the cervical spine, occipital pain, back ache, or pain radiating down the arm may be present. A radiograph of the cervical spine is indicated. Physiotherapy is most helpful.

THE EAR

Otalgia may be due to disease within the ear or be due to referred pain. The commonest causes of referred otalgia are impacted molars, tonsillitis (including tonsillectomy), dental infection, and temporomandibular dysfunction. Tumours of the oral cavity, pharynx or larynx may also result in ear ache and should always be considered.

THE EYE

Refractive errors and various ocular palsies may be associated with pain in the eyes and generalized headache. Glaucoma can also present with eye pain. Ethmoid and frontal sinusitis may cause deep seated pain around or above the eye.

THE SALIVARY GLANDS

Calculus disease and subsequent infection is relatively common in the submandibular gland but less so in the parotid. There will be a history of recent swelling and signs of inflammation. The duct of the gland can be palpated for a calculus. Likewise viral parotitis is accompanied by a systemic illness. Persistent pain in a salivary gland, particularly if a lump is palpable, is a sinister sign. The classic tumour to present in this way is the *adenoid cystic carcinoma* which invades via nerve sheaths.

FACIAL PAIN FROM A VASCULAR ORIGIN

Giant cell arteritis (temporal arteritis)

This disease is not restricted to the temporal artery, but is a generalized inflammation throughout the arteries in the head and neck. The most important artery it affects is the central artery of the retina. The patient may present with a unilateral blindness. There is usually generalized pain in the head, especially over the temporal region. The temporal artery itself is thickened and this can be felt by palpation. There are systemic signs with a raised temperature.

The erythrocyte sedimentation rate (ESR) is always raised. A biopsy of the temporal artery reveals the giant cell arteritis which obliterates the lumen.

Treatment with steroids is urgent if blindness is to be prevented.

Migraine

There are many variants of migraine. Typically, visual disturbance occurs, followed by a severe headache. Vomiting together with other motor and sensory signs may occur. There may be a family history. Migraine affecting the vertebrobasilar system may be accompanied by vertigo and ataxia. Treatment is with simple analgesics (paracetamol) and anti-emetics. Ergotamine derivatives are used in patients who do not respond to analgesics.

Cluster headache

The clinical syndrome consists of unilateral severe pain with a blocked nose. Typically the episodes occur in bouts, hence the name 'cluster' headache. The pain is intense and lasts for two to three hours. There is then a long period of remission before a similar bout of recurrent headache. Cluster headaches differ from classical migraine in that it is much more common in males. The pain also tends to be constant, usually in the region of one eye.

CAUSALGIA

This occurs when there has been injury to a nerve. The pain is typically a burning pain and it may be accompanied by atrophic changes in the skin.

ATYPICAL FACIAL PAIN

Not surprisingly, persistent facial pain undermines the psychological state of the patient. Atypical facial pain is a moderately common presentation of depression. The pain is usually severe and persistent and it may be associated with other features of depression. No physical signs of cranial nerve involvement can be detected. It can be a difficult problem to treat.

CHAPTER EIGHTEEN

Tumours and granulomas

....

INTRODUCTION

There is a wide histological variety of nasal and sinus tumours (Table 20). Most are benign with malignant nasal and sinus tumours accounting for less than 3 per cent of head and neck malignancy. Most nasal tumours present in a similar way and biopsy is required to exclude malignancy. *Always remember that a unilateral polyp or nasal mass needs removal for histological examination.*

Table 20. Histological classification of nasal and sinus tumours

Benign	Malignant
1. Epithelial tumours	
Papilloma	Squamous cell carcinoma
Adenoma	Adenocarcinoma
Inverting papilloma (Ringertz)	Melanoma
	Salivary tumours (e.g. adenoid cystic carcinoma)
2. Tumours of non-epithelial origin	
Osteoma	Osteogenic sarcoma
Ossifying fibroma	Fibrosarcoma
Angiofibroma	Angiosarcoma
Chondroma	Chondrosarcoma
	Lymphoma
	Rhabdomyosarcoma

BENIGN TUMOURS

Many benign tumours are readily apparent: papillomas (Figs 198 and 199), nasolabial cyst (Fig. 200), and rhinophyma (Figs 201 and 202).

Osteomas

Osteomas are the most common benign tumours and may be an incidental finding on a sinus radiograph (Fig. 203). They are most

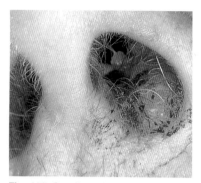

Fig. 198 Small papilloma situated at the nasal vestibule. It was removed by electrocautery.

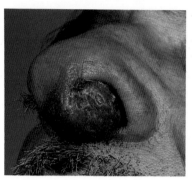

Fig. 199 This unsightly lesion was a large papilloma which was surgically excised.

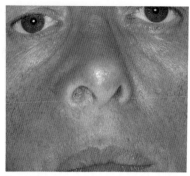

Fig. 200 Nasolabial cyst (right side). These cysts arise from fusion of embryological elements in the maxilla.

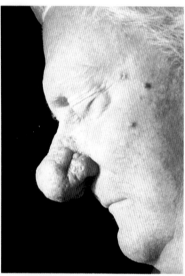

Fig. 201 Rhinophyma. This is not a neoplasm but is due to hyperplasia of sebaceous tissue in the nasal tip (see Fig. 202).

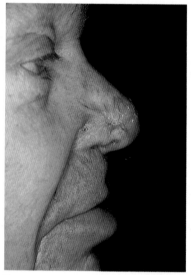

Fig. 202 Rhinophyma (the same patient as Fig. 201). The outcome of laser treatment.

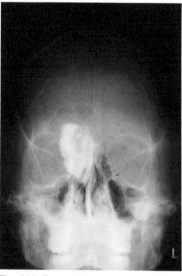

Fig. 203 Frontal sinus osteoma. Note the well demarcated calcified opacity with smooth edges (benign features).

common in the frontal sinus. The osteoma is usually asymptomatic but may cause headache and local tenderness over the sinus. Chronic sinusitis or a mucocele may develop if the ostium becomes obstructed. A frontal osteoma can displace the eye downwards and outwards. A well-demarcated calcified opacity with smooth edges (benign features) is seen on the plain radiograph. If an osteoma causes problems the treatment is complete removal. Large tumours may need craniofacial surgery.

Ossifying fibromas (fibrous dysplasia)

This condition of unknown aetiology affects the middle third of the face and the base of the skull in children and adolescents. One side of the face expands. The patient develops a cosmetic deformity and may have headaches, visual disturbance, and symptoms and signs due to cranial nerve compression. A large radio opaque mass arising from normal skeletal structures is the characteristic radiographic appearance. Cosmetic or other symptoms are treated by partial excision of the bony mass.

Inverting papilloma (Ringertz tumour)

The usual presentation is with a unilateral nasal polyp (Figs 204–206, p. 134). The papilloma arises from within the nose, the ethmoids, or the maxillary sinus. The patient has nasal obstruction and epistaxis is common. The tumour is benign but causes extensive local destruction. There is a small risk of malignant change. Radiological assessment with a CT scan is required. The treatment is removal of all the tumour via a lateral rhinotomy approach.

Angiofibroma

Pathology. This tumour occurs almost exclusively in adolescent males (age 10–25 years) and arises from the posterior nose (pterygoid plates). The proportions of angiomatous tissue and fibrous tissue vary. The tumour is benign but undergoes expansive growth.

Symptoms. Nasal obstruction, purulent nasal discharge, and frequent epistaxes. Obstruction of the Eustachian tube may cause a conductive deafness. Extension to the skull base leads to cranial nerve involvement.

Diagnosis. The diagnosis can often be made from the history. Frequent and heavy epistaxes in a young male are highly suspicious. If the tumour cannot be seen with anterior rhinoscopy it will be visible with the nasopharyngoscope.

Imaging. Plain lateral radiographs demonstrate a mass extending from the posterior nose into the postnasal space. Computerized tomography will demonstrate tumour spread and, if available, an MRI scan should give exceptionally good definition of the tumour. Angiography can demonstrate and enable embolization of the feeding vessels.

Treatment. The treatment of these tumours is controversial. The most usual treatment is surgical removal. Pre-operative embolization may reduce the operative bleeding. In some centres radiotherapy is used.

Key Point

Frequent heavy nose bleeds in an adolescent male should suggest a diagnosis of angiofibroma.

Fig. 204 Inverted papilloma (Ringertz tumour). This tumour was unilateral and does not have the gelatinous appearance of the typical nasal polyp (Fig. 205).

Fig. 205 Typical gelatinous appearance of the 'allergic' nasal polyp—contrast with Fig. 204.

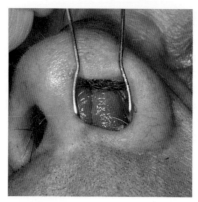
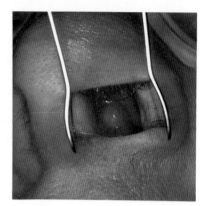

Fig. 204   Fig. 205

Fig. 206 Inverted papilloma—CT scan (axial). The tumour (T, white arrows) fills the maxillary antrum and the adjacent nasal cavity. Note the normal maxillary antrum (M).

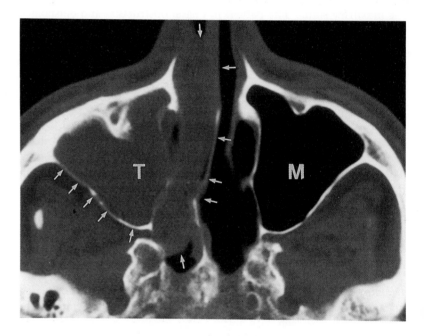

MALIGNANT TUMOURS OF THE EXTERNAL NOSE

Primary tumours of the external nose include basal cell carcinoma (BCC), squamous cell carcinoma (SCC), and malignant melanoma. These tumours are related to sun exposure. The head, and particularly the nose, receives a lot of sun. Of Caucasians living in the tropics of North Queensland, 85 per cent develop some form of skin cancer.

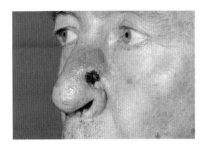

Fig. 207 Basal cell carcinoma. The rolled edges are typical.

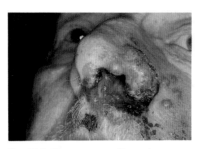

Fig. 208 Squamous cell carcinoma. Note the raised, everted edges (unlike Fig. 207).

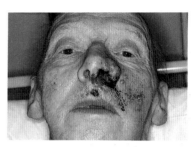

Fig. 209 Squamous cell carcinoma (the same patient as Fig. 208). The lump in the submental area was a lymph node metastasis.

Basal cell carcinoma (BCC)

Initially there is a firm nodule which increases in size. The growth rate is slow. The tumour then ulcerates (Fig. 207) and infiltrates into the underlying tissue. The degree of ulceration can be relatively small, but the extension into surrounding tissues can be considerable. Basal cell carcinomas usually do not metastasize. Their hallmark is gross local destruction. Troublesome areas are the columella and eye lids.

Treatment is by wide excision followed by immediate or delayed reconstruction. The prognosis of these tumours is good provided the tumour is excised with clear margins.

Squamous cell carcinoma (SCC)

This is a common tumour of the nose. Its growth rate is rapid and it ulcerates to form a crater (Figs 208 and 209). Regional lymph node metastases occur early. The treatment is total excision and this may be combined with radiotherapy. A radical neck dissection is required if there is neck node involvement.

Certain clinical features help distinguish between a BCC and an SCC. The former has a rolled edge and a slow growth rate but a definitive diagnosis can only be made by histology.

Malignant melanoma

This tumour is forty times more common in Caucasian populations living in the tropics than in those living in temperate zones. It can occur at any age. Ten per cent of malignant melanomas occur in the head and neck. The tumour may arise from a pre-existent pigmented mole, but this is not always the case. Suspicious symptoms include an increase in the size of a mole, change in the colour, and bleeding. Spread to the lymph nodes is common. The prognosis is related to the depth of

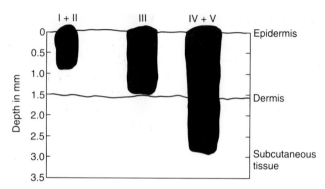

Fig. 210 Malignant melanoma. Prognosis is directly related to depth of invasion (Clark's levels).

spread and histologically is described according to Clark's levels (Fig. 210).

The mainstay of treatment is rapid and complete excision of the tumour. Nodal involvement may necessitate a radical neck dissection. Other therapies, including immunotherapy, chemotherapy, and radiotherapy have been employed.

MALIGNANT TUMOURS OF THE NASAL CAVITIES AND SINUSES

Introduction

Nasal and sinus malignancies make up only 3 per cent of head and neck tumours. A general practitioner may only see one in a lifetime. Sinuses are capacious and can harbour a large tumour silently. Basal and sinus malignancies have a poor prognosis with a 30 per cent 5 year survival.

Histology

A wide variety of histological types of tumour are encountered in the sinuses. The most common is squamous cell carcinoma (SCC), followed by adenocarcinoma, adenoid cystic carcinoma (and other salivary gland tumours), and malignant melanoma.

Aetiology

Only the adenocarcinoma has a known aetiology. An increased occurrence of the tumour was noted in High Wycombe furniture makers, particularly those using hard woods. The hard wood dust is inhaled, lodging in the ethmoidal mucosa.

Clinical features

A tumour arising in the nasal cavity is most likely to present with epistaxis and nasal blockage, whereas an ethmoidal tumour is more likely to present with orbital as well as nasal symptoms (Figs 211–217). Only 10 per cent of tumours have palpable neck nodes on first presentation. The lymph drains predominantly to the retropharyngeal and deep cervical nodes making palpation difficult.

Management

A careful history and examination followed by radiological examination particularly with computerized tomography (CT) is required for accurate diagnosis. CT is able to delineate tumour spread. Further information to distinguish soft tissue detail is provided by MRI (Figs 218 and 219).

Treatment

No single treatment is effective. Radiotherapy combined with surgery has produced the best results. While a good initial response to chemotherapy is often encouraging, it has not been shown to increase survival. The standard operations for sinus and nasal malignancy is maxillectomy. The orbital contents may need to be removed. If there is extension into the anterior cranial fossa both these approaches can be extended to include a craniofacial resection.

Dental prosthetics are able to add considerably to the quality of the patient's life by providing comfortable upper jaw prostheses allowing chewing and articulation.

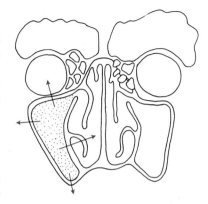

Fig. 211 Maxillary tumours may present with orbital, facial, dental, nasal, or neurological symptoms and signs (Figs 212–217).

Key Points

1. A unilateral nasal polyp or mass should always be biopsied.

2. Facial swelling is more often associated with tumour than sinusitis.

3. Wood workers are at risk of developing ethmoidal adenocarcinoma.

4. The most common tumour of the sinuses is squamous cell carcinoma followed by adenocarcinoma.

NASAL AND SINUS GRANULOMAS

Introduction

Many chronic inflammations are characterized by granuloma formation. These are divided into specific granulomas where a definite causative organism has been identified and non-specific where the aetiology is obscure (Table 21).

Fig. 212 Maxillary sinus carcinoma (right side). Note the swelling of the right cheek (see Figs 213 and 214).

Fig. 213 Maxillary sinus carcinoma (the same patient as in Fig. 212). Proptosis (and cheek swelling) are often better appreciated when the patient is viewed from above.

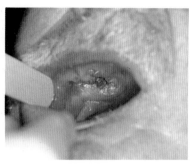

Fig. 214 Maxillary sinus carcinoma (the same patient as in Fig. 212). Examination of the oral cavity reveals that the tumour has breached through the hard palate.

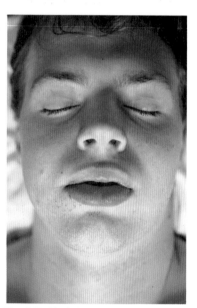

Fig. 215 Rhabdomyosarcoma maxillary antrum (right side). Note the cheek swelling (see Fig. 216).

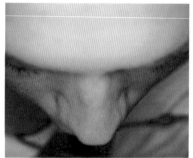

Fig. 216 Rhabdomyosarcoma maxillary antrum (right side). The cheek swelling is better appreciated when viewed from above (see Fig. 217).

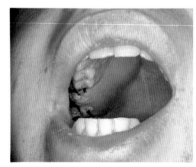

Fig. 217 Rhabdomyosarcoma maxillary antrum (right side). The patient had presented to his dentist with a swelling of the hard palate. An upper molar was extracted resulting in torrential haemorrhage (note the packing is still in place). The patient had limited jaw movement (trismus) due to backward extension of the tumour.

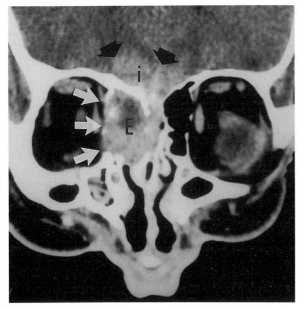

Fig. 218 CT scan (coronal). Right-sided ethmoidal carcinoma (E). note how well bone is visualized but the intracranial (i, dark arrows) and intraorbital (white arrows) extensions of the tumour are less clear than on MRI (Fig. 219).

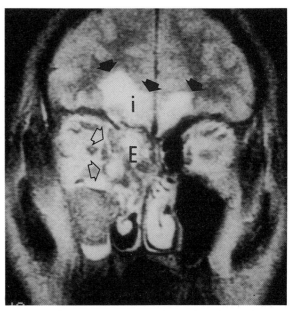

Fig. 219 MRI coronal scan showing a right-sided ethmoidal tumour (E) with intracranial (i, dark arrows) and intraorbital invasion (clear arrows). Note that bone is not imaged. Compare with the CT of the same patient in Fig. 218.

Table 21. Granulomas of the nose and sinuses

Specific granulomas	Non-specific granulomas
Tuberculosis	Wegener's granuloma
Syphilis	Malignant lymphoma
Leprosy	
Sarcoidosis	
Lupus vulgaris	
Aspergillosis	

Clinical features

Specific granulomas. Each type of specific granuloma has characteristic features but common to all is chronic granulations in the nose. The diagnosis is made by histology and microbiological techniques. Specific antimicrobial treatment is then given.

Fig. 220 Malignant lymphoma causing local destruction of the nose. This lesion was previously called 'lethal midline granuloma' or 'Stewart's granuloma'.

Non-specific granulomas: Wegener's granulomatosis. This is a systemic disease that is characterized by necrotizing granulomas in the respiratory tract and kidneys together with a generalized vasculitis. The patient may initially present to the ENT department feeling unwell and with a bloodstained nasal discharge.

Examination of the nasal cavity shows ulceration, granulation, crusting, and later a septal perforation may form.

The chest X-rays shows a diffuse infiltrative process. The erythrocyte sedimentation rate (ESR) is raised and the urine contains protein, red blood cells, and casts. Antineutrophil cytoplasmic antibody (ANCA) offers a specific diagnostic test and can be used to measure response to treatment. A raised creatinine confirms advanced renal involvement. Several biopsies may need to be taken before characteristic granulomas are seen.

The aim of treatment is to prevent renal failure. High doses of steroids and azathioprine are the mainstay of treatment. Cyclophosphamide is also used. The ESR level is used to monitor the disease activity.

Malignant lymphoma. This was previously referred to as a midline granuloma. Cell surface markers of malignant T cell lymphoma cells have now been demonstrated. The only similarity with Wegener's is that there is granulomatous disease in the nose (Fig. 220). There is a slow progressive destruction of the nasal and facial tissues and there is little systemic disturbance. The primary treatment is radiotherapy.

Key Points

1. The initial presentation of Wegener's granuloma may be to the ENT department with malaise, persistent nasal discharge, and granulation tissue in the nose.

2. Lesions previously called 'lethal midline granuloma' or 'Stewart's granuloma' were misnamed. These are definitely malignant lymphomas and are totally unrelated to Wegener's granuloma.

CHAPTER NINETEEN

Sleep disorders

INTRODUCTION

Sleep disorders need to be taken seriously. Severe sleep apnoea is a life threatening condition. ENT surgeons are increasingly involved in the management of sleep disorders as awareness of the problem increases. A multidisciplinary approach with close co-operation with respiratory physicians is necessary.

SNORING

It is estimated that between 40 and 60 per cent of men and 30 to 40 per cent of women snore. The tendency to snore increases with age. Snoring results from partial airway obstruction and it is the turbulent airflow in the upper airway that generates the noise. There is a spectrum from mild snoring to heroic snoring and sleep apnoea. A heroic snorer is one that can be heard from outside the room and quite often outside the house! Heroic snorers are pre-disposed to the sleep apnoea syndrome.

SLEEP APNOEA

The terminology is as follows.

1. *Apnoeic episode.* This occurs when no air passes the nose or mouth over a 10 second interval.
2. *Sleep apnoea syndrome.* This is defined as 30 or more apnoeic episodes occurring during a 7 hour sleep or when there are more than 5 episodes of sleep apnoea per hour.
3. *Central apnoea.* This is caused by lack of central drive from the respiratory centre. There is no air flow and no respiratory effort. An example of central apnoea is the respiratory depression caused by an overdose of opiate drugs.
4. *Obstructive apnoea.* This occurs when air is unable to pass the lips or nose despite a respiratory effort being made. A combination of central and obstructive apnoea can exist together.

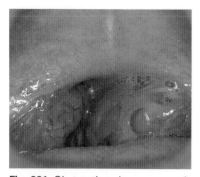

Fig. 221 Obstructive sleep apnoea. In this patient, the tonsils met in the midline. Adenotonsillectomy was curative.

The ENT surgeon is largely concerned with obstructive apnoea. In children the condition is caused by enlarged tonsils and occasionally adenoids (Fig. 221). In adults the condition is more complex and merits special investigation.

Pathophysiology

The snoring arises because of air turbulence around the soft palate. During sleep, muscle tone is reduced and the airway collapses on inspiration. Alcohol increases hypotonicity and exacerbates the problem.

Airway obstruction causes the arterial oxygen level to drop. The respiratory centre responds by increasing the inspiratory drive. Finally, when the respiratory effort is sufficient, the palate and pharyngeal tissues are sucked out of the way and the apnoeic period ends with a gasp.

The response of the pulmonary circulation to hypoxia is vasoconstriction and repeated hypoxic episodes lead to pulmonary hypertension. Hypoxia increases arrythmias and myocardial infarction may occur. Pickwickian syndrome describes gross obesity, sleep apnoea, and the tendency to sudden death during sleep.

History

The typical patient is an obese, middle-aged male who drinks heavily. The patient may complain of continual tiredness and may fall asleep during the day. One survey showed that one fifth of people with sleep apnoea syndrome admitted falling asleep while driving!

Close family or friends often give the best history. The spouse is only too well aware of the problem and usually has to sleep in a separate room. If apnoeic episodes are occurring the patient is often brought along because the sleeping partner thought they might suffocate!

Record in detail the person's alcohol consumption, smoking habits, exercise or lack of it, and weight gain.

Examination

Eighty per cent of heroic snorers are obese. Record height and weight. The classic anatomical deformity associated with snoring and sleep apnoea is a long soft palate, a large neck, and excess tissue around the tonsil. Associated abnormalities which can cause obstruction are a retrognathic mandible and a large tongue relative to the mouth.

Nasal deformity and blockage can contribute to snoring and need to be assessed. With the tip of the nasopharyngoscope in the post nasal space it is possible to assess the contribution to obstruction made by the soft palate and lateral pharyngeal walls.

The sleep study

A sleep study is an essential investigation because it allows the severity of the problem to be ascertained. It is possible to determine if the patient snores and if the snoring is accompanied by sleep apnoea. The most important parameters to measure are the arterial oxygen and respiratory movements.

Relatively small monitors are now available (Fig. 222) that accurately monitor the arterial oxygen, end tidal CO_2, and the ECG. The information gathered over an 8 hour sleep is stored and can be analysed at leisure. It should soon be practically possible for a patient to take such a monitor home so that he can sleep in familiar surroundings.

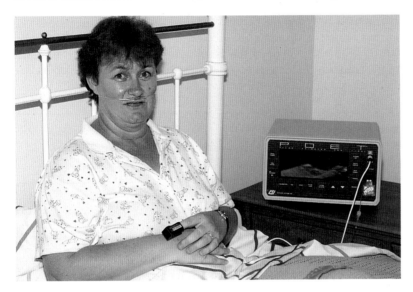

Fig. 222 'Sleep screening'. The finger probe is connected to a pulse oximeter which continuously records the arterial oxygen. The small catheter in the nose records the end tidal CO_2. ECG recording can be included. With this equipment it is possible to assess the severity of respiratory obstruction.

TREATMENT OPTIONS

Heroic snoring and or sleep apnoea is a serious problem and this fact needs to be conveyed to the patient and his family. Weight loss, regular exercise, and a reduced alcohol intake can alleviate the problem considerably.

Continous positive airway pressure (CPAP)

Over 300 anti-snoring devices have been patented; CPAP is one that works. It consists of a small mask connected to a pump that supplies positive pressure to the upper airway, thereby preventing collapse. CPAP will prevent sleep apnoea when the device is worn but it is not the most romantic device. Patients may comply with the device because they come to rely on the improved quality of sleep.

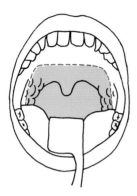

Fig. 223 The uvulo-palato-pharyngoplasty operation (UPPP) cures snoring and may help sleep apnoea.

Surgery

Some patients with sleep disorders are a poor anaesthetic risk and pre-operative anaesthetic assessment is prudent.

Tracheostomy

This is successful but a permanent hole in the neck is not without its problems.

Uvulopalatopharyngoplasty (UPPP)

In this operation redundant tissues in the soft palate and lateral pharynx are trimmed away and the tissues tightened (Fig. 223). If the problem is snoring alone the operation is successful (90 per cent cure). In individuals with sleep apnoea the results are less predictable. Specific complications include nasal regurgitation for the first few weeks following operation and occasional pharyngeal stenosis.

Key Points

1. Obstructive sleep apnoea occurs when there is no air entry despite respiratory effort.

2. Sleep apnoea may cause oxygen desaturation, pulmonary hypertension, and cardiac arrhythmias.

3. Heroic snoring and sleep apnoea are linked.

4. Surgery is more likely to cure snoring than sleep apnoea.

5. Weight loss and alcohol reduction are essential early measures in management.

6. Do not travel with an obese alcoholic at the wheel—they may fall asleep!

CHAPTER TWENTY

Pituitary disease

....

INTRODUCTION

The management of pituitary disease requires a team approach. Overall responsibility lies with the endocrinologist. Ear, nose and throat surgeons, neurosurgeons and radiotherapists may participate in the treatment. Surgery is the most commonly used method for treating pituitary tumours.

A pituitary tumour presents because of its local effects or its systemic effects.

LOCAL EFFECTS

A tumour extending upwards presses on the optic chiasma and may produce any type of visual field defect—the classic bitemporal hemianopia being occasionally demonstrable.

The contents of the cavernous sinus may be affected (cranial nerves III, IV, and VI). Raised intracranial pressure is rare with pituitary adenomas, and papilloedema is exceptional.

Downward extension into the sphenoid and postnasal space may present with CSF leak and recurrent meningitis.

SYSTEMIC EFFECTS

Acromegaly

Functioning adenomas secrete growth hormone and may be small (microadenoma) or replace most of the gland (Figs 224 and 225). High resolution CT scans can now demonstrate microadenomas and surgical removal leaving some normal gland behind is possible. Surgery is the most effective form of treatment for acromegaly.

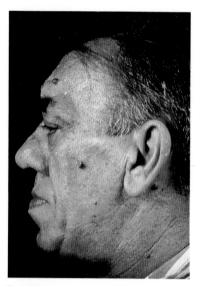

Fig. 224 Acromegaly. Note the coarse features of the leonine facies, simian ridges, and protuberant lower jaw. The referral letter asked 'does this patient have acromegaly or is he just ugly?'.

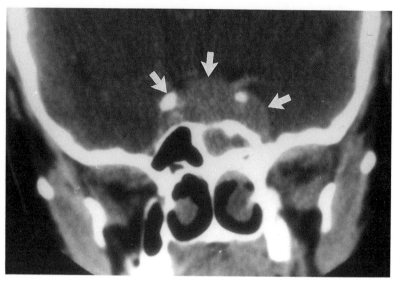

Fig. 225 Coronal CT scan showing a large pituitary tumour (arrows).

Cushing's disease

Cushing's can result from excessive steroid intake, an adrenal tumour, ectopic ACTH secreting tumours, and pituitary tumours that secrete ACTH. The pituitary tumours are often microadenomas that are amenable to surgery with excellent results.

Increased prolactin secretion

This can result from a tumour that secretes prolactin or from pressure effects in the region of the pituitary and hypothalamus. The condition presents with ammenorrhoea, infertility, and galactorrhoea in women and gynaecomastia in men.

Treatment with bromocriptine offers superior results to surgical micro-dissection.

SURGICAL APPROACHES TO THE PITUITARY GLAND

The pituitary gland can be approached in three ways.

1. From below, i.e., through the nasal cavity (trans-sphenoidal). This is the method of choice for excising the vast majority of tumours.
2. Through the ethmoid sinuses (trans-ethmoidal). This requires a small skin incision at the root of the nose.

3. Through the anterior cranial fossa (trans-frontal). This approach is needed only when there is extensive parasellar spread or the sphenoid anatomy is unsuitable.

Key Point

Any patient with progressive visual loss in whom the acuity cannot be corrected to 6/9 or better by refraction should be referred for neuro-opthalmological assessment to exclude a compressive lesion, such as a pituitary tumour.

CHAPTER TWENTY ONE

Paediatric rhinology

Table 22. Important causes of a nasal mass in childhood

> Nasal dermoid cysts
> Meningocoele
> Nasal glioma
> Olfactory neuroblastoma
> Rhabdomyosarcoma
> Nasopharyngeal carcinoma
> Lymphoma

THE MANAGEMENT OF A NASAL MASS

Many different entities may present as a nasal mass. It is important to remember that ethmoidal polyps are extremely rare under the age of two years. Pre-operative radiological investigation is of the utmost importance. The mass can be aspirated to check for the presence of CSF before a biopsy is taken. If there is no connection with the cranial cavity removal via the nose or a lateral rhinotomy is usually possible.

THE DIFFERENTIAL DIAGNOSIS OF A NASAL MASS

Nasal masses may present in childhood and pose a challenging diagnostic problem (Table 22).

Nasal dermoid cysts

These occur in the midline and are due to entrapment of epithelial cells (Fig. 226). A sinus may be present marked by a small dimple on to the skin. Occasionally the sinus track may extend up to the dura so that a pre-operative CT scan is essential.

Meningocoele

A meningocoele is a herniation of dura which includes CSF and brain tissue. It presents either as a soft tissue mass at the root of the nose or as an intranasal swelling looking like a polyp. CT scans are required to determine the extent of the meningocoele and the associated skull base defect.

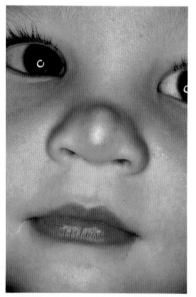

Fig. 226 Midline nasal dermoid. This small dermoid had no intracranial connection and was surgically removed from within the nose.

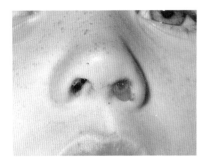

Fig. 227 Rhabdomyosarcoma of the nasopharynx. This child had a unilateral, smelly nasal discharge and was treated for months for 'sinusitis' (see Fig. 227).

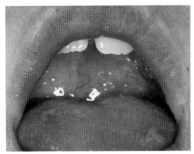

Fig. 228 Rhabdomyosarcoma of the nasopharynx. The soft palate was pushed downwards and the child sounded by hyponasal (see Fig. 229).

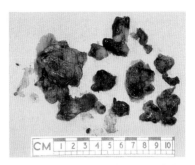

Fig. 229 Rhabdomyosarcoma of the nasopharynx. The appearances of the tumour at biopsy.

Nasal glioma

This is merely a solid mass of glial tissue that becomes entrapped outside the skull. The connection with the dura is usually lost and the cranium is intact. Most occur outside the nasal cavity and present as a lump on the nasal bridge but some are entirely intranasal.

Olfactory neuroblastoma (aesthesioblastoma)

This rare tumour develops from the olfactory placode and is related to neuroblastomas which develop from neural crest tissue. Nasal obstruction and epistaxis is associated with a mass in the nose.

Rhabdomyosarcomas

This is the second most common childhood malignancy. The tumour can develop in many sites in the head and neck including the sinuses, the nose, and the orbit (Figs 227–229). Treatment involves combinations of chemotherapy, radiotherapy, and surgery and should be undertaken in a paediatric oncology unit.

Lymphoma

This is the most common malignant tumour of childhood and can present as a mass arising from the back of the nose. Radiotherapy and chemotherapy now offer a chance of cure.

Nasopharyngeal carcinoma

This rarely presents in childhood (see p. 183).

Key Point

In a child with unilateral nasal symptoms exclude foreign body and tumour.

CHOANAL ATRESIA

This is a bony (90 per cent) or membranous (10 per cent) occlusion at the back of the nose (posterior choanae). The atresia may be bilateral or unilateral. A new born child is an obligate nasal breather, so bilateral choanal atresia is a fatal condition unless an oral airway is inserted.

Bilateral choanal atresia requires surgical intervention. The stenosis can usually be perforated with a probe and a silastic tube inserted on both sides through the choanae. A CT will give details of the extent of atresia prior to surgery. Repeated dilations may lead to an adequate opening or the atretic plate may need to be drilled out as an elective procedure.

Unilateral choanal atresia may present late. The child may have a unilateral nasal discharge or experience difficulty when the patent nostril is occluded during breast feeding.

> **Key Point**
>
> Bilateral choanal atresia presents at birth with inability to breathe.

SINUSITIS

The facial skeleton in a new born infant is dominated by the ethmoid sinuses which make up 50 per cent of the vertical dimension. Acute ethmoiditis develops rapidly and can have serious complications. Pus in the ethmoid sinus may erode through the lamina papyracea and lead to an orbital abscess.

The child presents with a history of pain around the eye (Fig. 230). The symptoms and signs develop rapidly and if the diagnosis is suspected urgent hospital admission is required. Proptosis and periorbital oedema are common. Limitation of eye movements and a reduction in visual acuity are sinister signs. If the condition does not respond rapidly to intravenous antibiotics or if there is any loss of visual acuity an urgent surgical exploration is required (see external ethmoidectomy, p. 105).

Fig. 230 Acute ethmoiditis. This young girl was febrile and complained of pain between the orbits and headache. There was slight swelling of her forehead. Intravenous antibiotics were curative.

> **Key Point**
>
> The ethmoid sinus is the dominant sinus in children. Ethmoiditis frequently presents with orbital complications.

FOREIGN BODIES IN THE NOSE

Beads, pieces of rubber, paper, and all sorts of oddments are frequently inserted into the nostril by children. They may not be detected initially but present with a unilateral purulent discharge. A good way of inspecting the nasal cavity in a child is to use the speculum of a fibreoptic

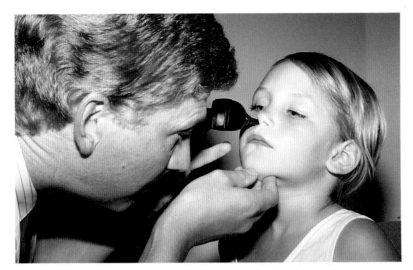

Fig. 231 The auriscope is an excellent instrument with which to inspect a child's nose, especially for a foreign body.

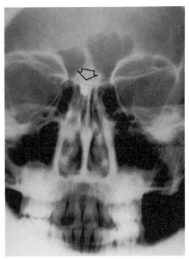

Fig. 232 This air gun pellet (arrow) lodged at the cribriform plate and was removed under anaesthesia.

auriscope (Fig. 231). This can be gently inserted without causing undue distress. If the child is co-operative, a foreign body in the anterior part of the nose can be removed by passing a probe under it and tipping it forwards into the front of the nose. Frequently a general anaesthetic is required to allow removal of the object (Fig. 232).

Key Points

1. Unilateral, offensive, purulent nasal discharge in a child is usually due to a foreign body.

2. A foreign body at the back of the nose is best referred to ENT for fear of dislodging it backwards into the tracheobronchial tree.

NASAL TRAUMA

The child's nose is less prominent and more cartilaginous than its adult counterpart. It is thus able to withstand the many facial injuries sustained by children. Be mindful of the possibility of non-accidental injury (Fig. 233). Injuries (surgery included!) may interfere with the growth centres in the nose with resultant cosmetic deformities in later life (Fig. 234).

Surgical correction of childhood nasal deformity should be conservative. Cartilage may be repositioned but not resected. Under many circumstances it may be better to delay surgery until the nose has reached adult proportions.

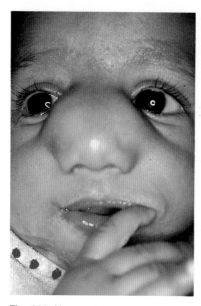 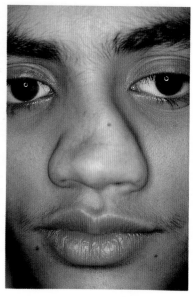

Fig. 233 Non-accidental injury. This grotesque injury was non-accidental. This diagnosis could easily have been overlooked.

Fig. 234 This ugly deformity resulted from an uncorrected childhood injury. The injury damaged the growth centres and the deformity increased with nasal growth.

SLEEP DISORDERS

Many children snore and a small proportion of these children have sleep apnoea (see p. 141). This is usually due to enlarged tonsils and occasionally to enlarged adenoids. Obstructive apnoea occurs when the tonsils totally block the airway so that no air enters despite strenuous respiratory efforts (sternal and intercostal recession can be seen). The low arterial oxygen then stimulates the respiratory centre sufficiently to produce a massive inspiratory effort which sucks the tonsils out of the way. The apnoea then ends.

Severe snoring and obstructive apnoea lead to secondary cor pulmonale and arrythmias. Tonsillectomy and/or adenoidectomy cures the problem. In children with severe obstruction it is wise to obtain a chest radiograph and an ECG. The anaesthetist should be informed in good time.

Key Point

Enlarged tonsils and adenoids may cause snoring and sleep apnoea in childhood.

CHAPTER TWENTY TWO

Cosmetic surgery

As specialists in the head and neck, it is not surprising that ENT surgeons are actively involved in aesthetic facial surgery. Patients can be greatly concerned about their facial appearance and identification of a cosmetic problem and its surgical correction can be both challenging and satisfying.

RHINOPLASTY

The appearance and function of the nose are closely related. Some patients may be reluctant to discuss cosmetic concerns. It is most important to encourage an open and frank discussion. The patient should be helped to point out their perceived problem and then be given a realistic appraisal of what cosmetic facial surgery can achieve.

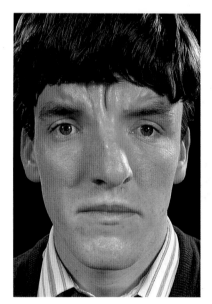

Fig. 235 The deviated nose. A septorhinoplasty is needed to restore function and appearance.

Fig. 236 A dorsal hump on the female nose is poorly tolerated. It is more acceptable on the male nose.

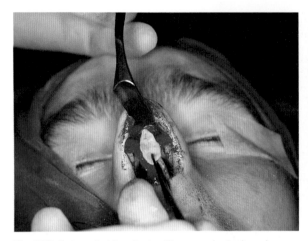

Fig. 237 External rhinoplasty. The nose is degloved allowing direct visualization of the nasal cartilages.

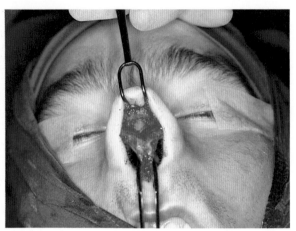

Fig. 238 External rhinoplasty. Here a cartilage graft is being positioned to augment the nasal dorsum.

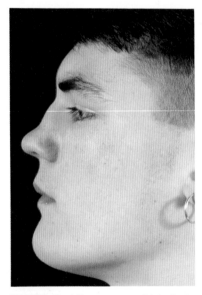

Fig. 239 Saddle nose pre-rhinoplasty (see Fig. 240).

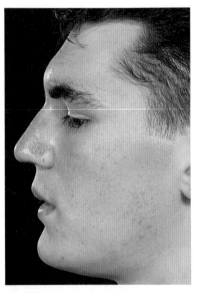

Fig. 240 The same patient as Fig. 239, post-augmentation rhinoplasty.

Key Point

A successful rhinoplasty should restore the normal structure and function of a nose. The reconstructed nose should harmonize with the rest of the face.

Abnormalities in the shape of the nose may be due to cartilaginous and/ or bony deformities. Some common abnormalities are shown in (Figs 235 and 236).

The basic requirements of rhinoplasty are that the surgery should restore normal function to the nose and the appearance should be improved so that the nose harmonizes with the rest of the face.

A deviated septum often occurs with a cosmetic deformity, so the standard operation is termed a septo-rhinoplasty (Figs 237–240).

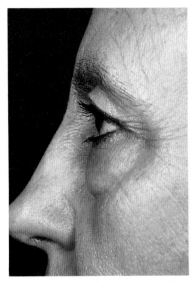 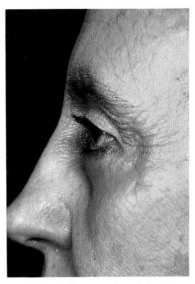

Fig. 241 (a) Pre-blepharoplasty. Note the 'bags' under the lower eyelid.

Fig. 241 (b) Same patient following blepharoplasty.

MENTOPLASTY

The nose should not be considered in isolation. A retrousse chin will make the nose appear large. Augmenting the chin will greatly improve the proportion of the face and may complement a rhinoplasty. Projection of the chin can be increased with a silastic implant or by bone advancement.

BLEPHAROPLASTY

Ageing may result in redundant skin and fat pads in the upper and lower eyelids (Figs 241 a and b). A blepharoplasty is an operation in which excess skin and fat are removed from the eyelids.

FACE LIFT

With increasing age the elasticity of the skin is lost (Figs 242 a and b). This process is accelerated by sun exposure. In the face lift operation incisions are made in hair line and the excess skin is pulled up and trimmed off. Wrinkles around the mouth and eyes can be treated by dermabrasion.

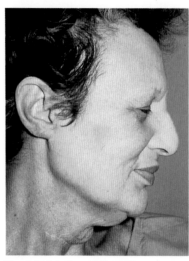

Fig. 242 (a) Face lifting. Note the pendulous appearance of the facial soft tissues.

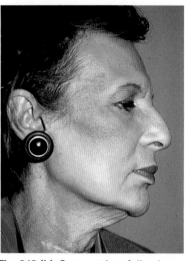

Fig. 242 (b) Same patient following face-lifting.

Fig. 243 Pinnaplasty. This patient was on a waiting list for surgery from childhood (see Fig. 244).

Fig. 244 Post pinnaplasty (see Fig. 243). The patient also demonstrates the aesthetic merits of a hair cut!

OTOPLASTY

Prominent ears (Figs 243 and 244) are best corrected before a child starts school. The usual deformity results from failure to develop an antihelical fold. The aim of surgery is to create a new natural-looking fold and various techniques exist to 'pin back' the protruding ears.

HAIR TRANSPLANTS

Flaps bearing hair may be swung from the occipital or temporal regions to cover the bald area of alternatively punch grafts containing several hair follicles can be applied.

Key Point

Careful patient selection and realistic expectations are essential in cosmetic facial surgery.

PART THREE

The Larynx, Head, and Neck

CHAPTER TWENTY THREE

Clinical anatomy and physiology

THE ORAL CAVITY AND PHARYNX

Many ENT diseases arise in this area and familiarity with the anatomy and physiology is important. The oral cavity contains the upper and lower dentition, the tongue and floor of the mouth, the hard palate and the openings of the major salivary glands (Figs 245 and 246). The area between the teeth and the cheeks (bucco-alveolar sulcus) is easily over-looked. The anterior border of the tonsil is known as the anterior pillar of the fauces and marks the start of the pharynx itself.

The tongue may be divided into an anterior two-thirds and a post-erior one-third. The anterior two-thirds comprises a dorsum, lateral borders and ventral surface. The posterior one-third of the tongue is continuous posteriorly with the epiglottis; between these two areas lie two small depressions known as valleculae which may be the site of food impaction.

The floor of the mouth is supported by the mylohyoid muscles which stretch between the rami of mandible like a hammock and are joined in the midline.

The oral cavity is lined by stratified squamous epithelium and it contains many small salivary glands ('minor' salivary glands). The blood supply is from branches of the external carotid artery. The lymphatic drainage is to ipsilateral nodes in the internal jugular chain but the anterior floor of mouth and base of tongue drain to both sides of the neck.

The *sensory nerve* supply of the tongue is from the lingual nerve in the anterior two-thirds and glossopharyngeal nerve (IX) posteriorly. The *motor supply* is from the hypoglossal nerve (XII). The *blood supply* is mainly from the lingual artery which is a direct branch of the external carotid artery.

THE TONSILS

This normally refers to the palatine tonsils which lie adjacent to the posterior one-third of the tongue but it should be remembered that there is a complete ring of lymphoid tissue (Waldeyer's ring) which

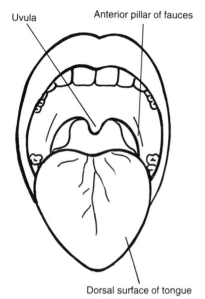

Uvula

Anterior pillar of fauces

Dorsal surface of tongue

Fig. 245 Diagrammatic representation of the oral cavity with tongue protruded.

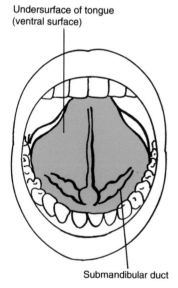

Undersurface of tongue (ventral surface)

Submandibular duct

Fig. 246 Diagram of the oral cavity with tongue touching hard palate and illustrating openings of submandibular ducts.

comprises the adenoids (pharyngeal tonsils), the palatine tonsils, and the lingual and pharyngeal tonsils. The latter are situated laterally on the posterior pharyngeal wall and are sometimes known as the lateral pharyngeal bands.

Immunological functions

The palatine tonsils are placed at the entry of the food and air passages and are constantly exposed to new antigens, unlike most of the lymphoid tissue in the body. They are part of the mucosal associated lymphoid tissues (MALT) and probably process antigen and present it to T helper cells and B cells. Although the majority of MALT cells produce IgA, the tonsillar tissue mainly produces IgG; IgA represents only about 35 per cent of its secretions. IgD is also seen but represents less than 5 per cent of secretions. These immunoglobulins pass directly out into the pharyngeal secretions and their output is enhanced in the presence of inflammation.

TASTE

The taste buds are made up of four different cell types which are found in the lingual epithelium. Microvilli project from the upper surface of these cells and are in contact with fluids in the mouth. In man, these cells lie in the mucosa of the oral cavity and pharynx but are also concentrated in the fungiform papillae in the anterior tongue and the circumvallate papillae which lie in an inverted V-shape at the junction of the anterior two-thirds and posterior one-third of the dorsum of the tongue (Fig. 247).

The taste sensations are transmitted to the tractus solitarius in the brainstem via two different routes.

Anterior two-thirds tongue

Afferent fibres travel from the tongue in the chorda tympani nerve which joins the facial nerve (VII) in the middle ear and passes into the posterior cranial fossa via the internal auditory meatus.

Posterior one-third tongue

Fibres pass via the glossopharyngeal nerve (IX) and enter the posterior fossa via the jugular foramen.

There are only four fundamental tastes; sweet, sour, bitter, and salt. Much of what we perceive as taste is really olfaction and patients who are anosmic often think they have lost their sense of taste and smell together. Accurate testing is very difficult due to the problems of standardizing stimuli.

SWALLOWING

This is a reflex which is mediated via *afferent* fibres passing to the medulla oblongata through the second division of the trigeminal (V), the glossopharyngeal (IX), and vagus (X) nerves. The *efferent* pathway is from the nucleus ambiguus and is mediated via the glossopharyngeal (IX), vagus (X), and hypoglossal (XII) nerves.

Oral phase

This comes first and involves the preparation of a food bolus in the oral cavity using teeth, tongue, and muscles of mastication. This is under voluntary control.

Reflex phase

This phase comes next and is initiated by displacement of the food bolus posteriorly by elevation of the tongue. Entry of the bolus into the oropharynx initiates the swallow and includes:

1. Inhibition of respiration and closure of the nasopharynx by elevation of the soft palate.

2. Elevation of the larynx under the tongue base.

3. Decrease of pressure in the lower pharynx and passage of the food bolus past the larynx into the two pyriform fossae towards the oesophageal inlet by a series of contractions of the pharyngeal constrictor muscles.

4. Relaxation of the cricopharyngeus muscle and peristalsis of the oesophageal musculature and transport of the bolus through to the stomach.

This process is sufficient to propel a food bolus even against the effects of gravity—thus enabling people to drink pints of beer while upside down!

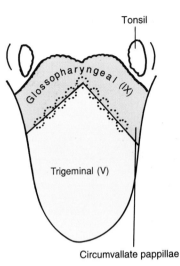

Fig. 247 Dorsum of tongue showing the V-shaped partition between the anterior 2/3 and the posterior 1/3, and the circumvallate papillae at the junction and the sensory nerve supply to the different parts.

THE LARYNX

The larynx is a protective sphincter at the inlet of the tracheo-bronchial tree and is also responsible for generation of sound. It has a mainly cartilaginous framework consisting of the hyoid bone above, the thyroid cartilage and cricoid cartilage below, and the arytenoid cartilages posteriorly. The cricoid cartilage is the only complete ring of cartilage in the entire airway. The larynx is situated in the anterior neck and its posterior wall is the same as the anterior wall of the pharynx. The epiglottis projects superiorly from the front of the larynx and acts like the lid of a box.

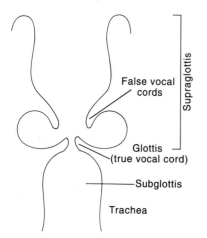

Fig. 248 Schematic vertical section through the larynx demonstrating the main anatomical and clinical sub-divisions.

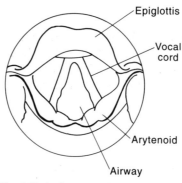

Fig. 249 A diagram of the larynx as seen during mirror laryngoscopy.

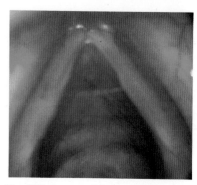

Fig. 250 A normal larynx as seen during fibreoptic laryngoscopy.

By convention the larynx is divided into supraglottis, glottis and subglottis. The true vocal cords (i.e. glottis) are white and contain the vocal ligaments which are at the medial end of the thyroarytenoid muscles. Above the muscles are two medial projections known as the false cords (Fig. 248).

The true vocal cords meet anteriorly at the approximate level of the thyroid prominence (Adam's apple). Posteriorly they are separate and each is attached to one arytenoid cartilage. This gives a V-shape in appearance (Figs 249 and 250).

The arytenoid cartilages articulate with the posterior part of the cricoid cartilage just below the glottis. The thyroid cartilage is also V-shaped and is open posteriorly.

Sensory nerve supply

The sensory nerve supply to the larynx above the vocal cords is from the superior laryngeal nerve. Below the vocal cords it is from the recurrent laryngeal nerve. Both of these nerves are branches of the vagus (X).

Motor nerve supply

The movements of the vocal cords and the intrinsic muscles of the larynx are controlled by the recurrent laryngeal nerve which is a branch of the vagus (X). The exception is the cricothyroid muscle which is supplied by the superior laryngeal nerve.

Key Point

Only the posterior crico-arytenoid muscles abduct the cords; all other intrinsic laryngeal muscles adduct the cords.

Lymphatics

The lymphatic drainage is different above and below the glottis, which acts as a watershed (i.e. just like the innervation). The lymphatic drainage of the supraglottis is to the upper deep cervical nodes while that of the subglottis is to nodes along the internal jugular vein and also to peritracheal and mediastinal nodes.

Key Points

1. The vocal ligament itself has no lymphatic drainage, hence tumours confined to the vocal ligament do not metastasize.

2. The glottis demarcates the neurovascular supply to the larynx. Above the cords, sensory innervation is from the superior laryngeal nerve. Below the cords it is from the recurrent laryngeal nerve.

3. All the intrinsic muscles of the larynx are supplied by the recurrent laryngeal nerve with the exception of the cricothyroid muscle.

Voice production

The lungs, diaphragm and abdomen provide a source of air. The larynx channels this into a column of high-speed vibrating air which is then converted into intelligible speech by the remainder of the vocal tract—i.e. the pharynx, tongue, lips, resonating chambers of the head, etc.

In brief, the larynx functions by closing the vocal cords against the air pressure from below which is created by exhalation. This causes a rise in subglottic pressure which forces the cords apart slightly for an instant. The subglottic pressure is thus reduced for a moment and the cords come together again (the Bernoulli effect). This occurs in rapid sequence to produce a vibrating stream of air emanating from the lungs.

Pitch is controlled by the frequency of the vocal cord vibrations which in turn are determined by the thickness, length, and tension of the cords. Intensity is governed by air pressure and the amplitude of vibrations.

THE NECK

The neck can be divided into triangles: the sternomastoid muscle separates the posterior triangle from the anterior triangle. The posterior triangle extends backward to the anterior border of trapezius and inferiorly to the clavicle. The anterior triangle extends to the midline and the upper part may be subdivided into the submandibular triangle (above the digastric muscle) and submental triangle (between the anterior bellies of the digastric) (Fig. 251).

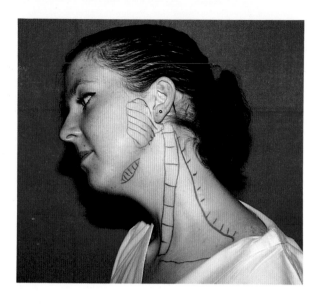

Fig. 251 Surface markings of the parotid and submandibular glands. The anterior border of the sternomastoid muscle is also identified as is the anterior rim of the trapezius muscle and the level of the clavicle.

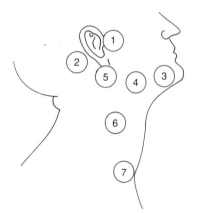

Fig. 252 Scheme of main lymph node groups in the neck: 1, preauricular; 2, occipital; 3, submental; 4, jugulodigastric (also known as the 'tonsillar' node); 5, 6, 7, upper, middle, and lower deep cervical.

The *lymphatic drainage* of the head and neck is of great clinical significance. While a small chain of nodes exists in the upper neck (submental, pre-auricular, occipital), the most important chain is the deep cervical nodes which run adjacent to the internal jugular vein. They are usually divided into four: the jugulodigastric ('tonsillar node') and the upper, middle, and lower deep cervical nodes (Fig. 252). When enlarged, they can be palpated along the anterior border of the sterno-mastoid muscle but they also lie underneath this muscle and their true size when enlarged can be very difficult to determine clinically.

THE SALIVARY GLANDS

Humans produce approximately 1 litre of saliva each day. The main salivary glands are pairs of structures—parotid, submandibular, and sublingual. These are known as the major glands.

The parotid gland is composed almost entirely of serous acini. The submandibular is a mixture of serous and mucinous acini (mainly serous) and the sublingual consists mainly of mucinous acini. There are also thousands of small collections of salivary tissue throughout the upper aero-digestive tract known as the minor salivary glands.

Saliva contains two enzymes—*ptyalin*, which comes from the major salivary glands and is an amylase, and *lingual lipase*. It also contains mucin which is a glycoprotein and acts as a lubricant fluid; some immunoglobulins including IgA, IgG, and IgD are also found. Saliva performs many functions including moistening, dissolving some molecules, facilitating speech articulation, and keeping the mouth clean by washing away bacteria around the teeth.

Secretion of saliva

The secretion of saliva is controlled by the *para-sympathetic secretomotor fibres* which synapse in ganglia near their end organ as follows.

1. *Parotid gland* via the glossopharyngeal nerve (IX). The tympanic branch sends fibres via the lesser superficial petrosal nerve to the otic ganglion. The post-ganglionic fibres then 'hitch-hike' with the auriculotemporal nerve which is a branch of the trigeminal (V).

2. *Submandibular gland* via the facial nerve (VII). The fibres pass to the submandibular ganglion in the chorda tympani.

3. *Sub-lingual gland* via the facial nerve (VII). This receives its secreto-motor supply in the same fashion as the submandibular gland.

The post-ganglionic fibres which enter the glands cause local release of acetylcholine or vasoactive intestinal peptide (VIP) which is a cotrans-mitter. This neurosecretion is initiated by the presence of fluid in the mouth or stomach. It can also be stimulated by the sight, smell, or thought of food—as in Pavlov's original experiments on the condition-ing of dogs.

THE THYROID GLAND

Anatomy

The thyroid is an endocrine gland which secretes tri-iodothyronine (T3), tetra-iodothyronine (T4/thyroxine), and calcitonin. It lies in the lower neck and weighs 20–30 g; it consists of two lateral lobes con-nected by a narrow midline isthmus which overlies the uppermost tracheal rings. It receives arterial blood from the superior thyroid artery—a branch of the external carotid—and the inferior thyroid artery which is a branch of the thyrocervical trunk. The posterior border is related to the oesophagus and the recurrent laryngeal nerves lie in the groove between the trachea and oesophagus, covered by the thyroid gland. These nerves are closely related to the inferior thyroid arteries. The gland consists of two types of secretory cells; *follicles* which are lined with epithelium and contain colloid and *parafollicular cells* which secrete calcitonin.

Function

The thyroid gland functions by trapping iodide from the blood which is coupled with tyrosine to form T3 and T4. The circulating levels of thyroid hormones are monitored by the hypothalamic–pituitary axis and *thyroid stimulating hormone* (TSH) is secreted from the anterior pituitary as required. This is a classical biological feedback mechanism.

The hormones themselves are stored as colloid within the follicles. The height of the epithelial cells in the follicles is controlled by TSH secretion. The parafollicular cells are of neural crest origin and are involved in calcium homeostasis.

Embryology

The gland develops in the floor of the pharynx from a thickening which forms the thyroglossal duct. This elongates and descends early in fetal development, prior to ossification of the hyoid bone, to reach its eventual site in the lower neck by the third fetal month. The duct then breaks up but a small pit on the dorsum of the posterior third of the tongue—the foramen caecum—remains to mark the origin of the duct. Abnormal thyroid tissue may be seen anywhere along the track of the thyroglossal duct but is almost always midline (p. 195).

THE PARATHYROIDS

Anatomy

The superior and inferior parathyroids are small (0.5 cm) glands derived from the fourth and third pharyngeal pouches respectively and are usually attached to the posterior part of the thyroid gland. Their exact position is very variable although the superior glands are about the size of a small pea and are related to the vascular pedicle on the postero-superior part of the lobes. The inferior pair of parathyroids are more variable in position and may occasionally be found in the superior mediastinum.

Function

The parathyroid glands secrete *parathormone* which maintains calcium and phosphorus homeostasis via a complex interaction with the gastro-intestinal tract, skeletal system, kidneys, and vitamin D. Loss of parathyroid tissue (e.g. after laryngectomy) leads to hypocalcaemia and *tetanic muscle spasm*.

CHAPTER TWENTY FOUR

Examination of the head and neck

THE ORAL CAVITY

Clinical examination requires a bright light and a tongue depressor—a right angled metal tongue depressor is very useful for this. ENT surgeons customarily use a reflecting mirror on the head or a headband-mounted fibre-optic light source which permits use of both hands to move the tongue and hold instruments. Always examine the bucco-alveolar sulcus and the floor of the mouth. Examination without a tongue depressor is usually inadequate (Figs 253 and 254). Generalized diseases may often be identified from examination of the oral cavity—e.g. the tongue fasciculation of motor neurone disease, the brown mucosal pigmentation of Addisons disease, the macroglossia of acromegaly, hypoglossal nerve palsy (Fig. 255), or tongue tie (Fig. 256).

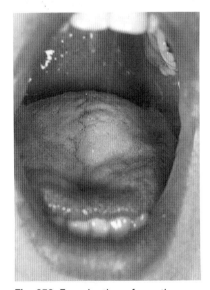

Fig. 253 Examination of mouth without a tongue depressor rarely reveals much information except the presence or absence of the tongue!

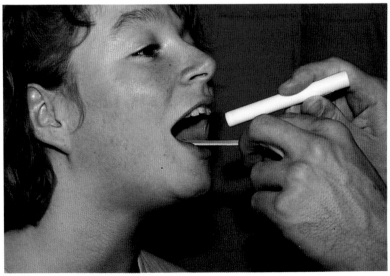

Fig. 254 Examination of the mouth in a co-operative adult with a pen torch and wooden tongue depressor.

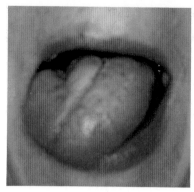

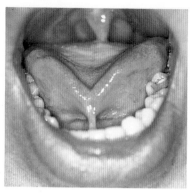

Fig. 255 A case of hypoglossal nerve palsy (XII). Note that the tongue always deviates to the side of the palsy.

Fig. 256 A case of a short lingual frenulum (tongue tie).

THE LARYNX, PHARYNX, AND NASOPHARYNX

Indirect laryngoscopy (Fig. 257)

This allows visualization of the larynx down to the level of the vocal cords. The tongue is protruded and held by the examiner while a warm mirror (Fig. 258) is placed on the posterior soft palate. In experienced hands it can be done rapidly and usually without any anaesthetic. It permits assessment of cord mobility as well as identification of mass lesions. The upper part of the pharynx and posterior part of the tongue are also seen with this technique. In some patients with a pronounced gag reflex, the procedure may be facilitated by use of a local anaesthetic spray such as lignocaine.

Fibre-optic laryngoscopy

This may be performed with a rigid or flexible endoscope (Fig. 259) and provides a clear view of all the structures involved. The fibre-optic nasendoscope is passed through the nose under topical anaesthesia (with cocaine) and the entire nasopharynx and larynx can be seen and demonstrated to others. It also permits photography for documentation of specific lesions (Figs 260 and 261).

Key Point

Thorough inspection of the larynx is essential in every hoarse patient.

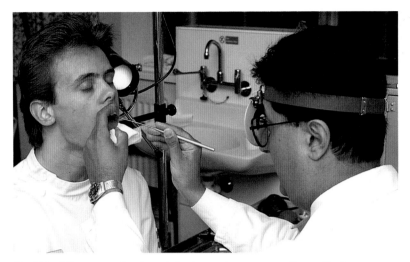

Fig. 257 Mirror examination of the larynx in a co-operative patient.

Fig. 258 A set of laryngeal mirrors as used for indirect laryngoscopy. These mirrors have to be warmed before insertion into the mouth.

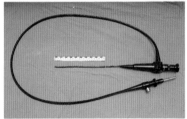

Fig. 259 Equipment for fibreoptic examination of the pharynx and larynx.

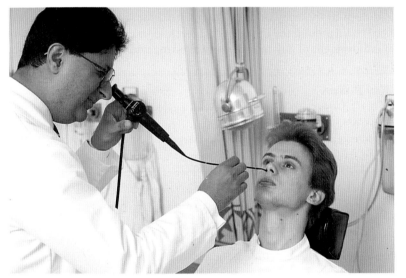

Fig. 260 Demonstration of fibreoptic assessment of the larynx with the nasendoscope in out-patients under local anaesthesia.

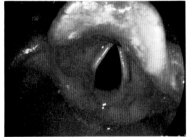

Fig. 261 Normal larynx photographed using a nasendoscope (see also Fig. 249).

THE NECK

Expose the whole neck and inspect the neck from in front. Then stand behind the patient and flex the chin downwards slightly to remove undue tension in the sternohyoid muscles and the platysma. Palpate the neck with the pulps of the fingers, not the tips (Fig. 262). Palpate the

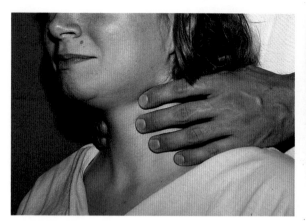

Fig. 262 Examination of the neck. Always stand behind the patient and use the pulps not the tips of the fingers. Keep the neck slightly flexed.

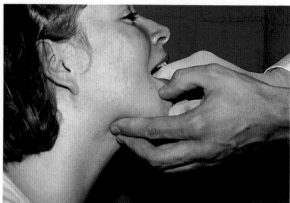

Fig. 263 Bimanual examination of the submandibular gland with a gloved finger inserted intra-orally.

deep chain of lymph nodes along the course of the interior jugular vein and deep to the belly of the sternomastoid muscle and also the superficial chain around the upper neck. Check for normal laryngeal mobility and movement of any masses on swallowing and on tongue protrusion.

Key Points

1. An enlarged neck node is often much bigger than it appears on palpation.

2. When examining for a lump in the neck it is helpful to ask the patient to locate it first.

THE SALIVARY GLANDS

Examine the area around the gland that appears to be affected and compare it with the other side. Always inspect the opening of the salivary ducts and examine regional nerves. This means the facial nerve for parotid lesions and the hypoglossal and lingual nerves for submandibular lesions. The gland should also be palpated bimanually with a gloved finger inside the mouth (Fig. 263). Do not let the patient open the mouth too wide as the surrounding muscles become tense and interfere with palpation.

Key Point

Never examine the parotid without testing the facial nerve.

The key points in the history and examination are summarized in Table 23.

Table 23. ENT: history-taking and examination

Ear	Nose	Throat
History		
Earache, irritation	Obstruction	Hoarseness
Deafness	Rhinorrhoea/post nasal drip	Dysphagia
Discharge	Allergy/hay fever	Stridor
Tinnitus	Facial pain	Lump in the neck
Vertigo	Epistaxis	
	Sense of smell	Sleep disturbance
	Appearance	
Children: Speech/language		
Past History		
Barotrauma	Trauma	Cigarette smoking
Acoustic trauma	Medications	Alcohol
Head injury	*prescribed*	
Ototoxics	*non prescribed*	
Family history	Previous surgery	
Previous ear surgery		
Examination		
Pinna	Shape	Mouth
Mastoid	Septum	Larynx—refer to ENT
Ear canal	Turbinates	Neck
Tympanic membrane	Airway	
Tuning forks	?Mucopus	
Conversational test	Facial tenderness	
Nystagmus	Facial sensation	
Facial nerve	Facial swelling	

CHAPTER TWENTY FIVE

Investigation of head and neck disease

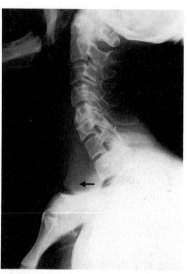

Fig. 264 Gross enlargement of the prevertebral soft tissue shadow in a case of carcinoma of the hypopharynx (arrowed).

THE ORAL CAVITY

Plain lateral X-rays

Plain lateral radiographs of the neck and cervical spine show soft tissue abnormalities. The depth of the prevertebral soft-tissue shadow can be a useful guide to the presence of disease in the pharynx (Fig. 264).

Orthopantomogram (OPT)

The OPT is a special radiological view which displays both upper and lower jaws and all the dentition on one X-ray plate (Fig. 265). It is helpful if searching for bone erosion or peridontal infection.

Intra-oral dental views

Intra-oral dental views are helpful especially when identifying salivary calculi (Fig. 266) (see also sialography, p. 178).

CT scanning

CT scanning has replaced tomography as the investigation of choice for many disorders and shows bone erosion and encroachment on to the great vessels of the neck by tumours (Fig. 267).

SWALLOWING

Barium swallow

Swallowing has traditionally been assessed by very simple methods such as the barium swallow. ENT surgeons have in the past supplemented the simple barium liquid studies with a solid object such as a biscuit or marshmallow coated with dye.

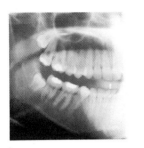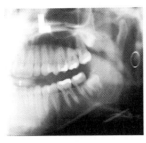

Fig. 265 A normal orthopantomogram in a young adult female.

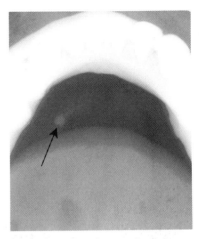

Fig. 266 Intra-oral x-ray view of a stone in the right submandibular duct.

Video fluoroscopy and high-speed cine recordings

Video fluoroscopy and high speed cine recordings have recently been developed and allow the evaluation of the oral and pharyngeal phases in much more detail. They are especially helpful in evaluating neurological disorders.

Manometric analysis and pH measurements

Manometric analysis and pH measurements are also used in specialized units and these can be coupled with simultaneous fluoroscopy.

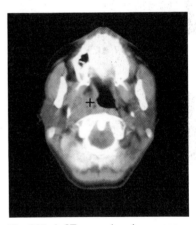

Fig. 267 A CT scan showing gross obstruction of the pharynx on the right side (marked by +).

THE LARYNX

Plain lateral X-rays

Plain lateral X-rays of the neck can give a little information—especially about the airway.

Tomography

Tomography can supplement plain films occasionally (Fig. 268) but is mainly used when access to CT scanning is not available.

CT scanning

CT scanning is the modern way of imaging the larynx and has superceded tomography. It is particularly good at demonstrating the size and local extent of malignant disease.

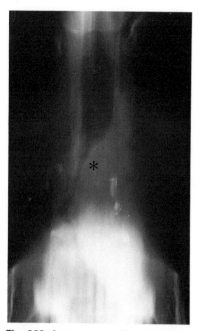

Fig. 268 A tomogram of the larynx demonstrating gross asymmetry of the airway due to a malignant tumour (marked *).

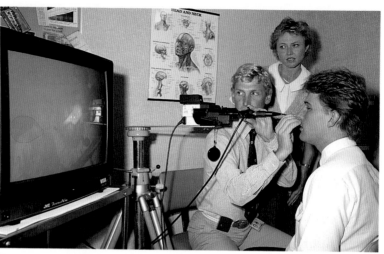

Fig. 269 A multidisciplinary voice clinic. Note the television screen upon which an image of the patient's larynx is being demonstrated to all the members of the team and to the patient himself.

Magnetic resonance imaging

Magnetic resonance imaging is being used increasingly often and represents another new and important option but it is not yet widely available.

Voice analysis

Several techniques are available for the measurement of voice including electrolaryngography and electromyography of the laryngeal muscles. The most promising is video-stroboscopy which can be undertaken as an out-patient technique with topical anaesthesia. This utilizes a flexible nasendoscope coupled to a video camera and a stroboscopic light which flashes on and off at the same speed as the object being viewed. This 'freezes' the image and enables the surgeon to identify the rhythm of the glottic movements and demonstrate this on a TV screen (Fig. 269) to the patient and speech therapist. This technique has allowed much more sensitive assessment of the mucosal waveform of the glottis.

Direct laryngoscopy and micro-laryngoscopy

Examination under general anaesthesia (EUA). This permits thorough examination of the larynx and biopsy, and may be carried out with the aid of the operating microscope (micro-laryngoscopy) which facilitates the photographic documentation of pathology. It is increasingly being replaced by fibre-optic examination (see above) but remains the main-stay of assessment in cases of suspected cancer. Micro-laryngoscopy also permits the specific treatment of certain disorders such as polyps.

THE NECK

Angiography or digital subtraction vascular imaging

This is essential if a vascular lesion is suspected and also has a therapeutic role by facilitating embolization.

CT scanning and magnetic resonance imaging

CT scanning or MRI are particularly helpful in imaging smaller lymph nodes which may be missed on clinical examination.

Fine needle aspiration cytology

Fine needle aspiration (FNA) cytology is very useful if a neck lump is thought to be malignant and this can be done in out-patients without anaesthetic. There is no evidence of spread of tumour through the skin track caused by the hypodermic needle used for FNA cytology (Fig. 270).

Ultrasound scanning

Ultrasound can be very informative in differentiating solid lesions (e.g. malignant lymph nodes) from cystic lesions (e.g. branchial cysts). This technique is also very helpful in the management of thyroid swellings.

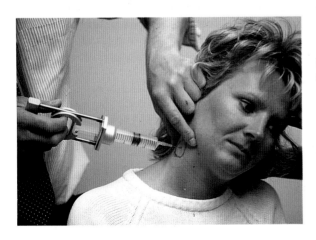

Fig. 270 A fine needle aspirate being taken in out-patients using a special sampling syringe.

Fig. 271 The stone that was removed from the patient in Fig. 266.

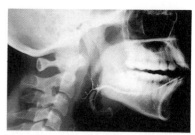

Fig. 272 A normal submandibular sialogram showing good ductal pattern.

THE SALIVARY GLANDS

Plain X-rays

Plain X-rays are of limited value except in the assessment of sub-mandibular calculi most of which are radio-opaque (Figs 266 and 271).

Sialography

The duct of the parotid or submandibular salivary glands can be cannulated and dye injected (Fig. 272). This displays the entire ductal system and can be helpful in demonstrating distortion of the ducts by tumour or chronic changes in the gland such as sialectasis.

CT scanning

CT scanning is of relatively limited value except when an obviously malignant lesion is being assessed for erosion of adjacent bony structures.

Magnetic resonance imaging

Magnetic resonance imaging is a superior technique for imaging salivary malignancies—especially those affecting the parotid gland (Fig. 273).

Fine needle aspiration cytology

Fine needle aspiration cytology offers advantages in the assessment of potential tumours as a few cells can be removed in out-patients for microscopic analysis without prejudicing the future management of the patient (unlike open biopsy). Interpretation of salivary cell smears is very difficult, however, and requires an exceptionally high level of expertise.

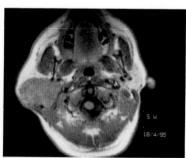

Fig. 273 A magnetic resonance scan of a large parotid tumour (bones do not show up on MRI scans).

The oral cavity and pharynx

ACUTE TONSILLITIS

This is usually a bacterial infection caused by a pyogenic group A Streptococcus. It is characterized by fever, sore throat, enlargement of the jugulo-digastric lymph node, and dysphagia. White pustules are seen on the palatine tonsils in the area corresponding to the lymphatic follicles—hence the name *follicular tonsillitis*. This acute inflammation is bilateral and almost always sensitive to benzyl or phenoxymethyl penicillin (penicillin V).

Tonsillectomy

Recurrent acute tonsillitis is the most common indication for tonsillectomy. It is usually considered when attacks have occurred a minimum of four times each year for at least 18 months. It is also advised for airway obstruction and the sleep apnoea syndrome, peritonsillar abscess, and in cases where malignancy is suspected.

Surgical points. Surgery on the upper part of the airway always carries significant risks—in particular from bleeding and inhalation. It is essential to bear in mind that the total blood volume of a child depends on its body weight. An approximate figure is 70–75 ml of blood per kilogram of weight (this figure is slightly higher in the neonatal period). From this it follows that a child of 15 kg has a circulating blood volume of just over 1 litre and that a loss of 100 ml certainly requires fluid replacement. Children who are bleeding post-operatively exhibit remarkable maintenance of their blood pressure until fairly late but the pulse rate always rises early on and is one of the most important clinical signs.

Peritonsillar abscess (Quinsy)

This is fully covered in Emergencies (p. 211).

> **Key Point**
> There is no evidence of deleterious long-term immunological side-effects from tonsillectomy.

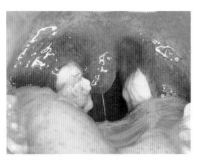

Fig. 274 The typical appearance of a creamy exudate covering the tonsils in glandular fever. This appearance is virtually diagnostic.

GLANDULAR FEVER

Aetiology

This systemic disorder is usually caused by the Epstein–Barr virus but a similar disorder can arise with other infective organisms such as toxoplasma or cytomegalovirus. A small proportion of patients with glandular fever syndrome have a predominantly pharyngeal disease.

Clinical presentation

The discomfort and dysphagia may be extreme with drooling of saliva and respiratory difficulty (usually inspiratory). The appearance of the throat is typical with a creamy-grey exudate covering both tonsils and appearing confluent (Fig. 274).

The pharyngeal appearances coupled with a high temperature are virtually diagnostic but the disease can be confirmed by serological testing which should show a positive Paul Bunnel test, an absolute and relative lymphocytosis, and the presence of atypical monocytes in the peripheral blood.

Treatment

Antibiotics are of little value except to prevent secondary bacterial infection—ampicillin should be avoided because of the frequent appearance of a papular rash. Steroids are occasionally required for short-term treatment if the airway is compromised. *If such treatment is necessary it should only be carried out in specialized units because intensive monitoring of the airway is essential.*

HUMAN IMMUNODEFICIENCY VIRUS

Aetiology

The human immunodeficiency virus which is the causative agent of acquired immune deficiency syndrome (AIDS) can affect almost all of the ENT system but it is most likely to be seen in the cervical region or in the oral cavity. This syndrome most often affects intravenous drug users and homosexual males, but it is spreading to affect the hetero-sexual community.

Clinical presentation

The clinical presentation of HIV infection in the head and neck is very varied but classically, these patients develop multiple oral vesicular ulcers especially due to herpetic infection. Any patient may present with an opportunistic infection such as *Candida* or pneumocystis but Kaposi's sarcoma may also arise in the oral cavity and presents as an elevated or flat erythematous nodule (typically on the palate) with intact overlying mucosa.

Cervical lymphadenopathy is also very frequent and this makes the evaluation of a neck mass in a high-risk patient very difficult. The neck nodes are usually enlarged due to follicular hyperplasia but lymphomas and other diseases are seen.

WHITE PATCHES

Any white patch which cannot easily be categorized as anything else is called *leukoplakia*. This usually refers to a dysplasia of the epithelial cells which is seen in smokers and is pre-malignant. Other causes of white patches include candida and aphthous ulcers.

Candida

Infection with *Candida albicans* produces confluent white areas which can be wiped off and are sensitive to antifungals. It is usually seen in patients who have had prolonged antibiotic treatment or who are immunosuppressed.

Aphthous ulcers

These are small painful ulcers 1–5 mm in size which may occur in crops throughout the oral cavity but are of unknown origin. They are associated with stress. They may be helped by steroid ointments and may recur over a period of many years.

THE GLOBUS SYNDROME

This is a description (Latin: globus = lump) for the feeling of a lump in the throat. It is commoner in women and mostly affects the 30–50 year age group. Dysphagia means difficulty in swallowing but these patients rarely have true dysphagia.

Aetiology and treatment

These patients have a combination of gastro-oesophageal reflux and spasm of the crico-pharyngeus muscle at the pharyngo-oesophageal junction. Other causes may include infected postnasal drip (due to sinusitis), pharyngeal pouch, and carcinoma of the pharynx. Most patients are treated by conservative measures such as anatacid therapy.

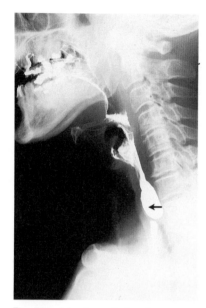

Fig. 275 A barium study of a pharyngeal pouch (arrowed) visible at the level of C7.

Fig. 276 Operative view of the pharyngeal pouch seen in Fig. 275 dissected from the neck and delivered externally prior to excision.

The original name of globus hystericus is unhelpful and should be abandoned, but psychological studies have shown that sufferers do show abnormal degrees of anxiety and introversion.

> **Key Points**
>
> **1.** The globus syndrome is a difficult diagnosis to make without full investigation.
>
> **2.** Patients presenting with a lump in the throat need careful evaluation to exclude local disease.

PHARYNGEAL POUCH

Aetiology

This is a false diverticulum arising at the junction of the oesophagus and pharynx. A true diverticulum contains all the layers of the organ of origin, but this lesion does not have the muscular layers present. The pharyngeal mucosa herniates posteriorly through a dehiscence between the upper and lower fibres of the inferior constrictor muscle. It typically occurs in males over 60 years old.

Clinical presentation

A pharyngeal pouch may present with a feeling of a lump in the throat (q.v.) and dysphagia with regurgitation of undigested food into the mouth some hours after eating. Chest infections and coughing are common due to overspill of the contents of the pouch into the airway, especially at night.

Examination may show some pooling of saliva on indirect or flexible laryngoscopy, but this is usually unremarkable. A barium study usually demonstrates the pouch (Fig. 275).

Treatment

This may be by an external approach to resect or invaginate the pouch (Fig. 276), or less commonly, by endoscopic resection of the partition between pharynx and pouch (most surgeons believe that division of the fibres of crico-pharyngeus is essential to prevent recurrence).

SIDEROPENIC DYSPHAGIA

This is an unusual combination of iron deficiency anaemia and dysphagia seen in middle aged women. It is also known as the Plummer–Vinson or Paterson–Brown–Kelly syndrome. It is characterized by atrophic mucosa in the pharynx and tongue, koilonychia, and dysphagia. The dysphagia is caused by a post-cricoid web and these patients have a high risk of post-cricoid malignancy (Fig. 277). This is thought to be because the cells in this area have one of the highest rates of iron turnover in the entire alimentary tract. The prognosis of post-cricoid tumours is appalling and radical tratment offers the only possibility for cure.

MALIGNANT TUMOURS OF THE MOUTH AND PHARYNX (INCLUDING NASOPHARYNX)

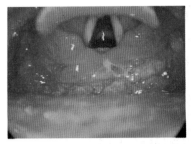

Fig. 277 Endoscopic view of a postcricoid carcinoma. The intimate relationship of this tumour to the larynx is clearly seen.

Aetiology

Tumours in the oral cavity and pharynx are relatively uncommon but it is important to bear in mind that the majority of these tumours can be identified clinically by an ENT specialist. Extraneous causative factors such as *cigarette smoking* and *consumption of spirits* are almost universal in this group of patients and are damaging to the entire mucosa of the aerodigestive tract; consequently these tumours do not always arise in isolation. A carcinoma may have extensive surrounding or adjacent mucosal dysplasia and these patients are more likely to develop a second primary malignancy within the aerodigestive tract (including the lungs).

Key Points

1. Almost all cancers in this region are squamous carcinomas.

2. Squamous carcinoma of the head and neck usually affects smokers over 50 years.

Nasopharynx

Carcinomas of the nasopharynx are commonly squamous although they have a tendency to demonstrate a pronounced lymphoid infiltrate. These tumours have a bimodal age distribution, affecting not only older people (over 50) like other head and neck cancers, but also younger patients in the second and third decades. This cancer is the commonest malignant tumour in southern Chinese.

Patients present with symptoms of *local disease, cranial nerve involvement at the skull base, or with neck metastates*. The *local symptoms* are nasal blockage and bloody discharge with a unilateral middle ear effusion. The *cranial nerves* comonly affected include the oculomotor nerves (III, IV, and VI), and the lower cranial nerves—IX–XII with trigeminal involvement. One third of these patients present with a lump in the neck—usually *upper deep cervical*.

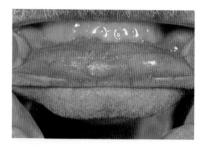

Fig. 278 A small squamous carcinoma of the lower lip in the midline which was successfully treated by radiotherapy.

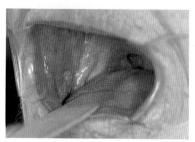

Fig. 279 A tumour on the right buccal mucosa in a smoker.

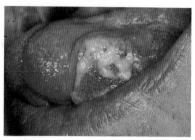

Fig. 280 A large carcinoma on the lateral border of the tongue treated by radical surgery.

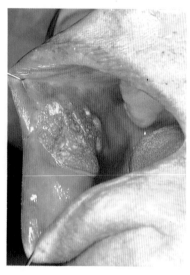

Fig. 281 A large squamous carcinoma of the right buccal mucosa prior to excision.

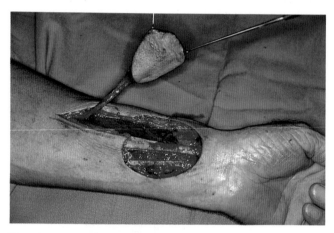

Fig. 282 A free flap being harvested from the patient's left forearm with attached radial artery and veins. The donor site is closed with a split skin graft.

Fig. 283 A close-up of the radial artery (on the right) about to be anastomosed to the facial artery (on the left).

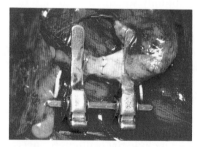

Fig. 284 A close-up view of the vessels seen in Fig. 283 after the sutures have been placed.

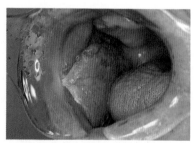

Fig. 285 Result of treatment in the patient from Fig. 281 one week after surgery. Note that the radial forearm skin now replaces the buccal mucosa.

These tumours can be diagnosed by flexible nasendoscopy and CT scanning and are usually treated by radiotherapy. Epstein–Barr surface antigen is a good serological marker of this disease.

Oral cavity

Oral cavity carcinomas are uncommon except in people who smoke and drink heavily or who chew tobacco or betel nuts (common in Asian communities).

They almost always present as *ulcerative painful lesions* (commonly on the lateral border of the tongue) (Figs 278, 279, and 280) with or without enlargement of regional lymph nodes.

Early tumours are well controlled by radiotherapy or surgery. Advanced carcinomas of the oral cavity and pharynx usually require combined tratment involving radical surgery with resection of tongue, mandible, and adjacent tissue. This type of radical treatment has been made easier in recent years with the advances in reconstructive techniques using skin, muscle, and bone flaps to repair the oral cavity. In this context, the development of microvascular anastamotic techniques in the past ten years has allowed skin flaps from distant sites such as the groin or forearm to be transplanted into the mouth and neck ('free flaps') (Figs 281–285). Advanced tumours are usually also treated with radiotherapy in a planned combined fashion with surgery.

Chemotherapy has no proven therapeutic role in the management of squamous carcinoma and its use is mainly confined to clinical trials.

Pharynx

Pharyngeal cancers (Fig. 286) are uncommon except in elderly male smokers. They commonly cause *painful dysphagia* and odynophagia (pain in the mouth on swallowing) and can usually be identified on clinical examination, especially with indirect laryngoscopy. They are often advanced at presentation and the overall outlook is poor.

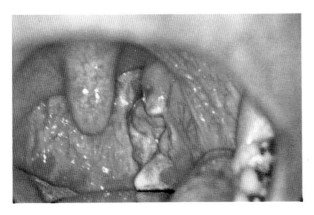

Fig. 286 Ulcerating carcinoma, left tonsillar region.

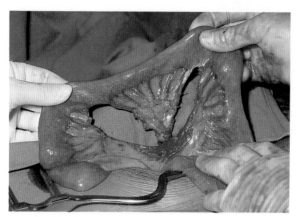
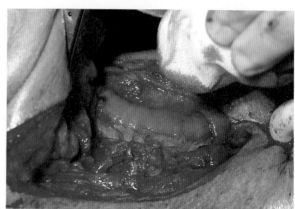

Fig. 287 Jejunum being mobilized prior to transfer to the neck in a patient with postcricoid carcinoma.

Fig. 288 Jejunum *in situ* in the neck. It is sutured to the tongue base above (on the left) and to the stomach below.

They may be managed by radiotherapy and radical surgery such as pharyngo-laryngo-oesophagectomy; this requires replacement of the pharynx and oesophagus by mobilizing the stomach ('stomach pull-up'), inserting a free jejunal graft (Figs 287 and 288), mobilizing the transverse colon, or converting a skin flap into a tube.

CHAPTER TWENTY SEVEN

Laryngeal disorders

Symptoms

There are three main symptoms of laryngeal disease: (1) hoarseness, (2) stridor, and (3) aspiration.

1. Hoarseness. This is primarily a symptom of laryngeal disease. Occasionally it is a manifestation of distant disease such as hypothyroidism or lung cancer.

2. Stridor. The larynx is the only part of the respiratory tract which contains an entire circle of cartilage—the cricoid cartilage. In addition, the area just below the vocal cords—the subglottis—is the narrowest part of the airway. Obstruction leads to diminished airflow and this leads to turbulence. The turbulent air flowing through a narrowed larynx causes a musical noise called stridor. This is usually *inspiratory stridor* when the blockage is at or above the level of the vocal cords and *biphasic stridor* if the blockage is in the subglottis or in the trachea. *Expiratory stridor* is usually called wheeze and arises because of a similar mechanism obstructing the small airways (e.g. in asthma).

3. Aspiration. This refers to the inhalation of food or saliva due to failure of the protective function of the larynx. It may manifest as coughing or choking when attempting to swallow and repeated chest infections due to saliva soiling the lungs. It is seen in patients with an immobile cord due to nerve palsy or tumour and in patients with a pharyngeal pouch.

Aspiration occurs more often with fluids rather than solids, i.e. the opposite of dysphagia due to oesophageal narrowing. This is because of the greater neuromuscular co-ordination required to swallow fluids.

> **Key Point**
>
> The larynx is a sphincter through which we breathe and which protects the airway during swallowing. It also acts as the generator of speech.

Signs

There are four main signs of laryngeal disorder.

1. Voice abnormalities. To the experienced specialist abnormalities of voice quality may be quite characteristic—e.g. the breathy voice of a vocal cord palsy or the harsh voice of chronic laryngitis. To the lay person, almost any term may be applied to encompass all abnormalities!

Dysphonia: merely a description of abnormal voice and not a diagnosis.
Laryngitis: This is a description of a specific appearance of the larynx
and cannot be diagnosed without viewing it.

2. Stridor. The noisy breathing of the stridulous patient is easily
detected. This is more common in young children where the airway is
smaller (p. 206). Since the subglottis is the narrowest part of the airway,
a small change here may have a disproportionately large effect on
airflow.

> **Key Point**
>
> Airflow is proportional to the fourth power of the radius of a tube
> (Poiseuille's formula) so a small reduction in size of the airway (e.g. at the
> level of the larynx) may have a large effect on airflow.

3. Mobility. Occasionally the mobility of the larynx to palpation
may be impaired due to tumour, but this is usually a late sign.

4. Lump in neck. Carcinoma of the larynx metastasizes to the neck
and this may be one of the presenting signs in advanced cases.

> **Key Point**
>
> Laryngitis is not an alternative term for hoarseness.

Disorders of the larynx are classified in Table 24.

Table 24. Causes of hoarseness

Specific voice disorders
  Mass abnormalities, e.g. nodules,
    polyps
  Diffuse abnormalities, e.g. acute
    and chronic laryngitis
  Vocal cord palsy
  Laryngeal cancer

Non-specific voice disorders
  Voice strain
  Functional dysphonia

SPECIFIC VOICE DISORDERS

Mass abnormalities

Vocal nodules. These are also known as singer's or screamers nodules
(Fig. 289). They are symmetrical small white swellings on the apposing
surface of the true cords, usually situated at the junction of the anterior
one-third and posterior two-thirds. They are the result of poor voice
production or chronic overuse—therefore speech therapy is the best
treatment. Occasionally they will need to be surgically removed using
micro-surgical techniques or endolaryngeal laser surgery followed by
speech therapy to correct the underlying errors in voice production.

Papillomata. Respiratory papillomata are rare in adults. They usually
occur in children and they tend to spread throughout the trachea-

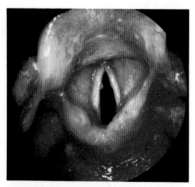

Fig. 289 Endoscopic view of small
vocal nodules (singers' nodules) seen
at the junction of the anterior 1/3 and
posterior 2/3 of the vocal cords.

bronchial tree airway if not carefully treated. They are of viral origin and are removed by laser surgery coupled with anti-viral agents such as interferon. The value of endoscopic laser surgery is that it causes very little scarring and does not appear to facilitate the implantation of the papillomata further down the airway. It is a great advance over earlier, more crude techniques.

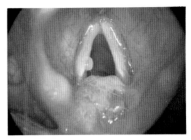

Fig. 290 A polyp on the posterior part of the left vocal cord.

Polyps (Fig. 290). Polyps are uncommon and often unilateral. They tend to follow an acute infective episode and will not resolve spontaneously. They require removal by micro-surgery or laser treatment.

Diffuse abnormalities

Acute laryngitis. This occurs as part of a diffuse upper respiratory infection or as a localized disorder. It is often viral and settles quickly. Active treatment is not essential but steam inhalations are soothing. It does not last more than three weeks in the vast majority of cases.

Key Point

Hoarseness lasting more than three weeks should always be referred for an ENT opinion in case it is due to an early laryngeal cancer.

Chronic laryngitis. Chronic laryngitis may be *specific* or *non-specific*. This separates inflammations due to a known agent such as tuberculosis, leprosy, etc. (i.e. chronic specific) from those due to an unknown agent.

Chronic specific laryngitis. The identification of laryngitis due to mycobacteria, syphilis, fungi etc is by biopsy. These disorders are now rare in the developed world. Treatment is directed towards the causative organism, e.g. dapsone for leprosy.

Chronic non-specific laryngitis. The main predisposing factors are smoking, chronic upper respiratory sepsis (chronic sinusitis with post-nasal drip), and chronic lower respiratory sepsis (chronic obstructive airways disease). In some cases, the laryngeal mucosa may become *dysplastic*—especially over the surface of the true vocal cords. This may be a pre-malignant condition—hence the need for expert appraisal.

The condition is treated by elimination of any predisposing factors such as smoking and by regular stripping of the affected areas of the vocal cords by microsurgery or by laser surgery using the CO_2 laser (Fig. 291).

The risk of malignant change depends on the severity of any dysplasia and continuation of the provoking factors.

Key Point

The diagnosis of chronic laryngitis should never be made unless the larynx has been fully assessed by a laryngologist.

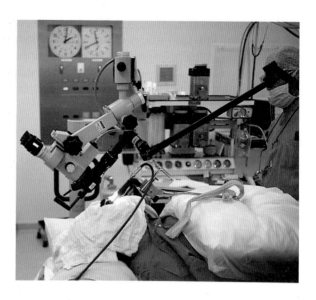

Fig. 291 Equipment set up for micro-laser surgery to the larynx in a case of laryngeal dysplasia.

Vocal cord palsy

This may be unilateral or bilateral. Unilateral is commonest and left cord palsy is the most frequently encountered.

Unilateral. Unilateral left vocal cord palsy is the commonest because the left recurrent laryngeal nerve pursues a long intra-thoracic course arching round the aorta, and is commonly involved in neoplasms involving the left hilum. Since lung cancer is the commonest single cancer, it must be considered first. Other lesions can cause a similar effect including cancer of the thyroid, oesophagus, and nasopharynx. Neurological lesions should also come into the differential diagnosis (see Table 25).

Bilateral. Bilateral cord palsies are uncommon and tend to occur after thyroid surgery or head injuries.

Key Points

1. Left vocal cord palsy is usually considered due to a carcinoma of the lung until proved otherwise.

2. Although cancer of the lung is the commonest single cause of left cord palsy, it is not the commonest cause of hoarseness.

Table 25. Causes of vocal cord palsy

Cause	Cord	Comment
Iatrogenic	Either	May follow thyroid, or chest surgery
Bronchial cancer	Left	Common
Skull base lesions	Either	
Thyroid/oesophageal cancer	Either	
Cardiomegaly	Left	Esp left atrium
Neurological diseases	Either	Multiple sclerosis
Idiopathic	Usually left	Common. Often follows viral infection

Treatment. The larynx often compensates in cases of unilateral palsy; if no improvement in voice quality occurs after three to six months, improvements can be achieved by laryngeal surgery techniques which are known collectively as phonosurgery. *Unilateral palsies* may be managed by injecting Teflon paste or bovine collagen into the affected side to bring it closer to the midline. This may be injected directly into the vocal fold under local or general anaesthesia. Alternatively a part of the laryngeal framework and its associated soft tissues can be brought surgically nearer the midline. This has the advantage of being reversible—unlike intracordal injection of Teflon. In *bilateral palsies*, the cords are often close to the midline (paramedian position), and therefore the airway is already compromised. These patients therefore often require tracheostomy. Surgery may be carried out on one arytenoid cartilage—either to move it to a more lateral position or to excise it altogether with the CO_2 laser as an endoscopic procedure. Some surgeons advocate reinnervation procedures for the paralysed larynx—e.g. rotating in a nerve-muscle combination from the neck. Any strap muscle can be used but the omohyoid appears to be the most useful. However, these techniques have not yet found widespread acceptance.

Key Points

1. In bilateral cord palsy there is always a 'trade-off' between improving the airway and restoring voice quality.

2. Idiopathic cord palsy can only be diagnosed after all other causes have been excluded.

Laryngeal cancer

Carcinomas of the larynx are predominantly squamous carcinomas and this is one of the commonest head and neck cancers, almost always occurring in elderly male smokers. Over the past 20 years the sex incidence has changed and 20 per cent of these tumours now occur in women. As in the oral cavity and pharynx, the normal stratified squamous epithelium of the larynx can exhibit a range of changes from mild dysplasia to carcinoma in situ to invasive carcinoma (Fig. 292). The majority of laryngeal carcinomas (55%) arise from the vocal cords—i.e. glottic. About 40 per cent arise from the supraglottis and a small proportion are primarily subglottic.

Presentation. Because of the sites of origin, patients almost always present with *hoarseness* and this is of great importance because early cancers have a 90 per cent 5-year cure rate. The cure rate drops dramatically with increased size of tumour at presentation.

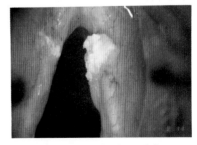

Fig. 292 Endoscopic view of the larynx with marked dysplasia and keratosis on the anterior part of the right vocal cord.

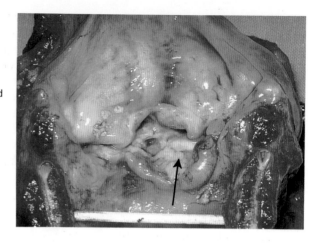

Fig. 293 A typical laryngectomy specimen photographed close up and demonstrating a large ulcerative tumour at the level of the true cord (arrowed). The specimen has been split in the midline posteriorly and is being propped open by a small wooden stick at the lower margin of the picture.

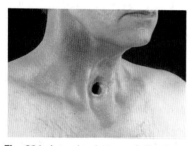

Fig. 294 A tracheal stoma following total laryngectomy. The patient's lungs have no other connection to the outside world.

Treatment. The vast majority are treated primarily by radiotherapy for cure with surgery usually reserved for recurrent or advanced tumours. Laryngeal carcinomas tend to spread via lymphatics to the cervical nodes and their involvement is a poor prognostic sign.

Laryngectomy. This radical operation for advanced or recurrent cancer involves a permanent interruption of the connection between the upper aerodigestive tract (nose, mouth) and lower respiratory tract (lungs). Part or all of the thyroid gland is usually removed at the same time (Fig. 293) and the patient breathes through a tracheal stump which is sutured to the neck skin (Fig. 294). After laryngectomy, the patient cannot immerse his neck in water—even to have a bath or shower. Well motivated patients, however, can overcome even this problem (Fig. 295).

Thyroxine and calcium supplements need to be taken orally for life if the whole thyroid gland has been removed.

Key Points

Remember to check thyroxine and calcium levels regularly in patients who have undergone laryngectomy.

Vocal rehabilitation. The loss of the larynx (vocal generator) does not prevent speech production as long as an alternative source of vibrating air can be found in the pharynx. Three alternatives are available: (1) artificial devices which cause vibration of air in the oral cavity or pharynx—usually battery-powered; (2) voice production may be restored in some patients by regurgitating air into the pharynx and then using this as a source for articulation (*oesophageal speech*); (3) a small channel can be created in the tracheal stoma to allow exhaled air to pass into the pharynx via a one-way valve (tracheo-oesophageal speech). This involves the implantation of a small device such as a Blom–Singer valve (Fig. 296).

NON-SPECIFIC VOICE DISORDERS

Voice strain

This is quite common in people who use their voices a lot. It is associated with incomplete closure of the vocal cords at some point along their length. A version of this also occurs in some patients who are on treatment with inhaled steroids when the cords present a 'bowed' appearance on attempted phonation.

Treatment is by voice rest and speech therapy.

Functional dysphonia

Functional dysphonia usually covers all conditions where there is no obvious disorder and the larynx appears to be moving normally. There may be a number of problems underlying the dysphonia which are not immediately apparent, ranging from excess muscle tension in the strap muscles of the neck to stress, emotional conflict, personality disorder or frank psychiatric illness. It is commonly seen in young women. Speech therapy is the most useful treatment, although a psychologist may be required in some cases.

Fig. 295 A rather complex device being worn by a laryngectomy sufferer to enable him to go swimming. Note that the patient has managed to reconnect his nasal and oral airway with his trachea.

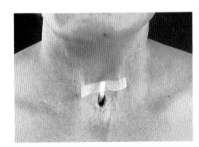

Fig. 296 A similar patient to Fig. 294 but with a small Blom–Singer valve in the stoma being held in place by a piece of white tape.

CHAPTER TWENTY EIGHT

Neck, salivary glands, and thyroid gland disease

LUMPS IN THE NECK: GENERAL COMMENTS

The correct diagnosis of a lump in the neck requires a careful history and examination. Most clinicians will gain 90 per cent of their diagnostic information from this alone.

Site, size, duration, and consistency of the mass are essential items of information. Midline swellings are usually congenital dermoids (Figs 297 and 298) or of thyroid origin. Masses along the line of the sterno-mastoid muscle are usually either lymph nodes or branchial cysts (see Table 26). Fluctuant painful swellings are usually due to infected abscesses. Remember the surgical sieve: i.e. congenital or acquired. If acquired, is it infectious, inflammatory, endocrine, or neoplastic?

Table 26. Causes of lumps in the neck

Midline	Lateral
Thyroid anomalies	Lymph nodes
Dermoid cysts	Branchial cysts

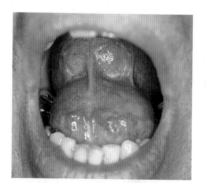

Fig. 297 A large midline dermoid visible in the floor of the mouth.

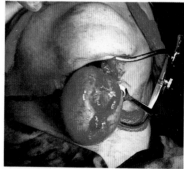

Fig. 298 The dermoid from Fig. 297 being delivered externally after an incision below the chin.

BRANCHIAL CYST

This is a malformation which usually presents in the upper neck in early or middle adulthood. A swelling is detected at the junction of the upper third and the middle third of the sternomastoid muscle on its anterior border (Fig. 299). On initial presentation there may be an acute illness due to infection of the cyst and the diagnosis can be confirmed by ultrasound. Treatment is by excision. An acutely inflamed cyst should never be incised as this will convert it into a branchial sinus, which is much more difficult to excise adequately.

THYROGLOSSAL DUCT CYSTS

Embryology

The thyroid gland descends early in foetal life from the base of the tongue towards its eventual position in the lower neck. At the time of descent, the hyoid has not been formed and the track of descent may pass in front, through, or behind the eventual position of the body of the hyoid. Thyroglossal duct cysts represent persistence of this track and may be found anywhere in the midline from the tongue base to the lower neck. Caution needs to be exercised as these abnormalities are sometimes the only functioning thyroid tissue in the body.

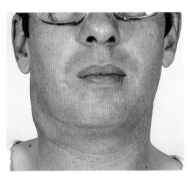

Fig. 299 A branchial cyst in the right upper neck.

Presentation

Cysts almost always arise in the midline (Fig. 300) and may also present at a time of acute infection. The classical description is of a cyst which moves upwards on swallowing and tongue protrusion, but this merely confirms attachment to the hyoid bone. These cysts may also rupture onto the skin of the neck and present as a discharging sinus.

Treatment

Excision is the best treatment, but the body of the hyoid bone and the suprahyoid track must be excised in order to prevent recurrence (Sistrunk operation).

LYMPH NODES

Small rubbery lymph nodes are typical of local infection, especially in children, but large nodes (more than 3 cm) or multiple nodes are highly suggestive of malignancy. A lymphoma is common in young adults but squamous carcinoma is common in older patients (over 50 years) (Fig. 301). In view of this, excision biopsy is not appropriate until a full ENT examination has been performed to identify the primary tumour (see Table 27). Ideally this should be done under general anaesthetic. If no tumour is evident, biopsy of the nasopharynx and pharynx, and tonsillectomy should be performed as this will often reveal a submucosal primary carcinoma. Treatment of squamous carcinoma of the neck is usually by a radical neck dissection rather than radiotherapy.

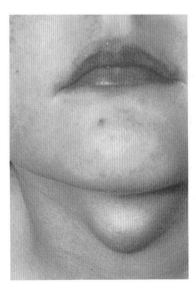

Fig. 300 A typical midline thyroglossal cyst.

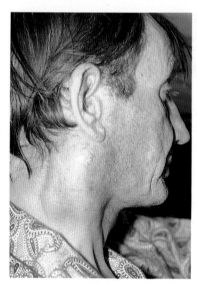

Fig. 301 A large right upper neck lump (upper deep cervical) in a patient with a carcinoma of the tonsil (see also Fig. 252).

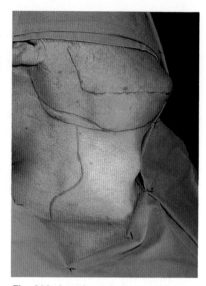

Fig. 302 A patient about to undergo radical neck dissection. The incision demonstrates a small ellipse of skin which needs to be excised following a previous incision biopsy. This man was of southern Chinese origin and had a carcinoma of the nasopharynx.

Table 27. Disadvantages of biopsy of neck lumps without ENT assessment

1. The patient may feel the lump has been removed and not attend for further follow-up
2. The neck has a scar which complicates future clinical assessment (Fig. 302)
3. Tumour seeding may have occurred into the neck tissues and the skin
4. The pathology report of 'metastatic squamous carcinoma of unknown origin' is of no value to the patient or clinician and represents a wasted general anaesthetic

---

Radical neck dissection

This is the basic operation for all malignant neck disease and was first described in 1906. The operation is standardized for all tumours except papillary carcinoma of the thyroid (p. 204), malignant melanoma, and lymphoma. The principle is to remove all of the main lymph node bearing tissues of the neck, either alone, or *en bloc* with a primary tumour. It usually also leads to removal of the sternomastoid muscle, internal jugular vein and accessory nerve (XI), in addition to the lymphatic tissue (see Table 28). Some modifications of this standard operation have been proposed over the last 30 years, but it continues to be the mainstay of treatment for malignant neck disease.

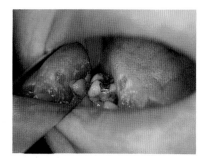

Fig. 303 A trophic ulcer on the right side of the tongue following damage to the lingual nerve during radical neck dissection. This patient has anaesthesia over that side of the tongue.

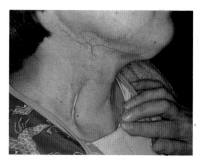

Fig. 304 Unsightly web formation in the lower part of a scar following radical neck dissection. Vertical skin incisions in the neck should always be curved in order to avoid this (see also Fig. 302).

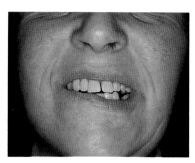

Fig. 305 A paralysis of the right lower lip following radical neck dissection and damage to the lower branches of the facial nerve.

Radical neck dissection is a major procedure and may be complicated by injury to the facial, vagus, glossopharyngeal, and hypoglossal cranial nerves in the upper neck and the brachial plexus and phrenic nerves and thoracic duct in the lower neck (Figs 303, 304, and 305).

Table 28. Parts of normal anatomy affected by a radical neck dissection

Excised	Retained
Internal jugular vein	Carotid artery
Accessory nerve	Vagus nerve
Sternomastoid muscle	Hypoglossal nerve
Submandibular gland	Brachial plexus
	Phrenic nerve

Key Point

A lump in the neck in an adult is due to a tumour arising in the pharynx or larynx until proven otherwise.

OCCULT PRIMARY CANCER

The primary site of cancer may remain unidentified in a proportion of patients with malignant squamous carcinoma causing lymph node enlargement in the neck, despite thorough investigation and biopsy as above. This is known as an occult primary and treatment usually involves radical neck surgery with or without radiotherapy. Long-term follow-up is required, but one-third of patients may never have the primary site identified.

CAROTID BODY TUMOURS

These rare lesions arise from chemoreceptor tissue at the bifurcation of the common carotid artery at the level of the hyoid bone and tend to grow upwards towards the skull base.

Clinical presentation

These are usually benign and give rise to a pulsatile swelling in the upper neck at the anterior border of the sternomastoid muscle which may involve adjacent lower cranial nerves. The swellings are classically mobile from side to side but not up and down, and can sometimes be emptied by firm pressure, after which they will refill in a pulsatile manner. There should also be an associated overlying bruit. Carotid body tumours may also present as a parapharyngeal mass displacing the tonsil medially. Angiography is essential and surgical excision is the treatment of choice.

Key Point

CT scanning is an essential investigation for a parapharyngeal lesion of unknown origin.

NEUROGENIC TUMOURS

These are rare in the neck but may arise from the vagus or glosspharyngeal nerves. They may present as lumps in the neck or as parapharyngeal masses and require excision after full investigation.

NECK SPACE INFECTIONS

These are dealt with in Chapter 29.

SALIVARY GLAND DISEASE: GENERAL COMMENTS

History

Salivary glands typically become enlarged when affected by disease— usually due to the accumulation of saliva which cannot flow normally into the oral cavity. Pain accompanying such swelling is not uncommon. Swelling which is intermittent and related to eating suggests intermittent obstruction such as a stone. Swelling which remains constant or slowly increases in size suggests tumour.

Examination

Clinical inspection and palpation of the ducts is very helpful. The parotid ducts open into the buccal mucosa at a small papilla opposite the second upper pre-molar. The submandibular ducts open into the floor of the mouth anteriorly at a small papilla near the midline and the frenulum of the tongue (Fig. 306).

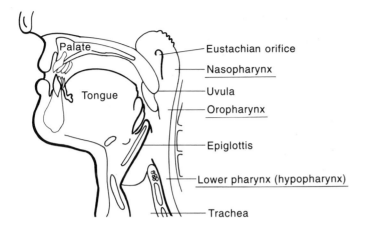

ACUTE PAROTITIS

This usually affects only one gland at a time and causes swelling, erythema, and pain. It may be caused by an acute bacterial infection or by a virus (e.g. in mumps, which is usually bilateral and mainly affects the parotid). It is also seen in patients who are generally debilitated or dehydrated, e.g. post-operative or following stroke. Treatment is by rehydration and intravenous antibiotics.

SALIVARY CALCULI (SIALOLITHIASIS)

These may affect any of the major glands but are commonest in the submandibular gland. They cause a sudden colic-like pain in the area of the gland in association with eating. If the salivary outflow is obstructed, the whole gland may swell up.

The diagnosis is usually not difficult as most calculi are radio-opaque. If the stone is in the duct, it may be removed via an intra-oral incision (Fig. 307). Recurrence is common and leads to irreversible changes in the gland itself which usually necessitate removal of the whole gland.

Fig. 307 A large calculus removed from a submandibular duct.

Key Point

Salivary calculi most commonly arise in the submandibular gland or duct.

SIALORRHOEA

Some patients—especially mentally handicapped children—suffer from drooling of saliva. A full ENT assessment is required and surgery such as adenoidectomy may help. In some cases, the submandibular ducts need to be transposed posteriorly to lie near the tonsil and this often has a very beneficial effect.

BENIGN SALIVARY TUMOURS

Pleomorphic adenoma

This is the commonest tumour of salivary tissue and the most frequent tumour in the parotid gland; it is also known as a mixed parotid tumour. Pathologically it consists of epithelial and myoepithelial cells (hence the old name of 'mixed tumour'). Pleomorphic adenomata account for 70 per cent of parotid masses and approximately 40 per cent of submandibular swellings. They are slightly commoner in women and the maximum incidence is at 40–50 years.

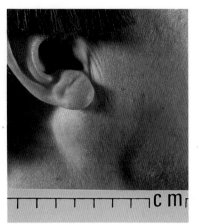

Fig. 308 A swelling over the angle of the right mandible in a patient with a pleomorphic adenoma of the parotid gland.

Clinical presentation. These lesions usually present as a lump which for years has been slowly enlarging, situated over or just behind the angle of the mandible (Fig. 308). Pain or other symptoms are rare. The swelling is firm, non-tender, and not fixed to overlying structures/mucosa. No facial nerve signs are found.

Treatment. Treatment is impossible without histological assessment. Accordingly the correct treatment is excision biopsy of all these lumps (basic surgical principle). In the parotid region this usually involves superficial parotidectomy with facial nerve preservation. Recurrences are nodular, multicentric, and usually fixed and are said to be much less likely if formal superficial parotidectomy is performed at the time of the original operation.

Excision of recurrences involves greater risk of surgical damage to the facial nerve than primary surgery.

Postoperative radiotherapy (RT). Radiotherapy may be valuable if a minor salivary gland tumour has been incompletely removed or if tumour seeding has occurred per-operatively. The principle of using radiotherapy for benign disease is contentious as the long-term effects of RT also include degeneration of normal tissue into malignancy (e.g. RT-induced sarcomas).

Superficial parotidectomy vs. lumpectomy. Some surgeons have advocated simple excision of the parotid lump ('lumpectomy'). The basic surgical principles which should always be observed are demonstration of the facial nerve and removal of the tumour mass with a cuff of 'normal' tissue. These are best accomplished by superficial parotidectomy.

Key Point

All lumps in the parotid region are tumours of the parotid gland until proven otherwise.

Warthin's tumour

Clinical presentation. This is also known as an adenolymphoma. It is commoner in men over the age of 60 (male:female = 5:1) and is occasionally bilateral (less than 10 per cent). Warthin's tumours represent one-tenth of parotid swellings and this is the most commonly affected gland—the lower pole of the gland is usually affected and has a typical cystic feel and ovoid shape. Symptoms are rare and these tumours usually present as slowly growing asymptomatic lumps.

Treatment is as for any parotid swelling. These tumours are encapsulated and have an excellent prognosis.

Haemangioma

These usually occur in children under 1 year. They are commonest in the parotid gland and most regress spontaneously so are not treated unless actually enlarging. If excised, they rarely recur.

Lymphangioma

These are similar to haemangiomas and they usually occur under the age of 2 years and in the parotid. They should be treated conservatively.

MALIGNANT SALIVARY GLAND TUMOURS

Adenoid cystic carcinoma

This is the commonest malignancy of salivary tissue and has a very characteristic clinical pattern. It has a predilection for neural tissue and hence nerve palsy and pain are common early symptoms.

Clinical presentation. The age of onset is 20–60 years, and they are more common in women. Adenoid cystic carcinoma represents the commonest malignancy in the submandibular or minor salivary glands and 30 per cent of all salivary malignancies. The predilection for neural spread causes early neuropathy and pain which is out of proportion to the size of the lesion. They tend to recur after many years (5–15 years) partly because of their tendency to grow proximally up nerves. They also have a tendency for blood-borne spread to the lungs (cannon ball metastases).

Treatment. These tumours are not radio-sensitive and are usually treated by primary radical surgery. The natural history is of a slowly but inevitably recurring malignancy. Aggressive treatment is justified but ultimate cure is very unlikely.

Adenocarcinoma

These form 3 per cent of parotid tumours and 10 per cent of extra-parotid salivary tumours. They usually arise after the age of 30 and occur equally in both sexes. They are also occasionally seen in children. These tumours are best treated by *radical local excision* as they are not radio-sensitive, but distant metastases occur.

Squamous carcinoma

These are rare in salivary tissue and are twice as common in men as women. They usually occur in elderly patients (over 60 years) and grow rapidly, involving nerves, skin (Fig. 309) and regional lymph nodes. The prognosis is poor.

Treatment consists of radical excision and radiotherapy.

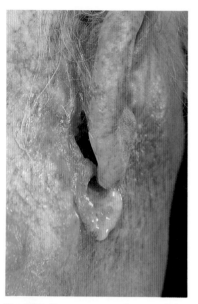

Fig. 309 Squamous carcinoma of the parotid causing skin breakdown and fistulization following unsuccessful primary radiotherapy.

Mucoepidermoid tumour

These make up 5–10 per cent salivary tumours and 90 per cent of them arise in the parotid. They may occur at any age, including childhood. Mucoepidermoid tumours are slow growing with a tendency to local recurrence, but regional lymph-node spread is seen in up to 30 per cent of cases and distant spread in 15 per cent. There is a distinct histological variation: low grade tumours have a 90 per cent 5-year survival rate and high grade tumours 30 per cent 5 year survival.

Treatment is usually by local excision but radiotherapy is used as an adjunct for high grade lesions.

Key Point

A parotid swelling with a facial palsy is usually neoplastic.

Lymphoma

Three-quarters of salivary lymphomas are found in the parotid gland. There is usually a short history (less than six months). Pain and facial palsy are uncommon. As with all lymphomas, clinical stage and histological sub-type determine survival. Median survival is four years.

Key Point

In children mucoepidemoid carcinomas and pleomorphic adenomas are the commonest tumours but vascular lesions predominate in infancy.

THYROID DISEASE

General comments

The term *goitre* just means enlargement of the gland and does not predict the disease. However, it has commonly been associated with the enlargement found in geographical areas where iodine is in short supply and hence with hypothyroidism (e.g. Derbyshire).

Hypothyroidism

This may be *primary* hypothyroidism due to abnormal uptake of iodine, synthesis of abnormal thyroxine, or due to absence of the gland (e.g. after surgical resection with laryngectomy). *Secondary* hypothyroidism occurs when the pituitary fails to produce TSH to stimulate the thyroid gland. The symptoms include cold intolerance, drowsiness, and a hoarse voice.

Hyperthyroidism

This may be due to primary hyperthyroidism or due to a toxic adenoma. The symptoms include irritability, insomnia, sweating, palpitations, etc. Such patients usually have a tachycardia and may develop eye signs such as lid lag and exophthalmos which may require surgical decompression through the maxillary antrum. The mainstay of treatment is with anti-thyroid drugs such as carbimazole.

Solitary thyroid nodule

Clinical presentation. Most patients will have no features of biochemical thyroid disease but care should also be taken to elicit any history of neck irradiation in childhood as the thyroid gland is known to be very susceptible to malignant change even 20–30 years later. In addition, a family history of thyroid cancer raises the possibility of medullary carcinoma (see below) and multiple endocrine neoplasia syndrome (MEN) which is an inherited condition.

Management. The critical question for any solitary thyroid lump is whether or not it is really part of a multinodular goitre (always benign) with the other nodules remaining clinically impalpable. This question is answered by *ultrasound* examination. This has the added advantage of demonstrating if a genuinely solitary nodule is cystic or solid. Cystic nodules can then be aspirated for cytology. Solid nodules can be aspirated or excised.

Radio-isotope scanning with technetium 99m (Tc 99m) can be used instead of ultrasound to determine whether a nodule is hot or cold (i.e. takes up radio-isotope or not). Hot nodules are said to be usually benign but cold nodules to have a risk of malignancy of 8–20 per cent. However, where fine needle aspiration is practised, isotope scanning can be reserved for patients who are toxic (hyperthyroid) or those in whom metastases are being sought for treatment with radio-iodine.

MALIGNANT THYROID DISEASE

Papillary carcinoma

Clinical presentation. This is the commonest tumour of the thyroid gland and may represent up to 60 per cent of thyroid cancers. It is usually seen in young adults and typically spreads to cervical lymph nodes (40 per cent). This disease in the lymph nodes was, at one time, known as 'lateral aberrant thyroid' because of the apparently well-differentiated nature of the tumours. Other metastases are found rarely—pulmonary in less than 5 per cent of patients.

These tumours may be under TSH control. *Histologically* some follicular elements are usually found, but all tumours with any papillary features are treated as papillary even if they are mainly follicular.

Treatment. If a single nodule, it is usually treated by lobectomy with or without the administration of radio-iodine I^{131}. If lymph nodes are present, total thyroidectomy and conservative excision of lymph nodes is indicated. A radical neck dissection is not indicated in this disease.

Follicular carcinoma

Clinical presentation. These represent 25 per cent of thryoid carcinomas and occur in older people (20–50 years). They typically spread by blood borne emboli especially to bone and lung (cannon-ball metastases). *Histologically* the tumours mimic typical glandular acini. Nodal spread is rare (less than 5 per cent).

Treatment. Usually by total thyroidectomy. Following excision, give TSH to stimulate any metastases and then give radioiodine (I^{131}).

Medullary carcinoma

Clinical presentation. This is an uncommon tumour of neural crest origin and arises from the parafollicular or C cells. These cells are part of a 'family' of cells throughout the body known as Amine Precursor Uptake and Decarboxylase cells (APUD). In the thyroid they are

---

Key Point

The more follicular a tumour the better its differentiation, and therefore the more likely it is to take up radioiodine.

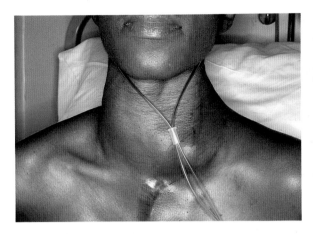

Fig. 310 A huge neck swelling caused by an anaplastic carcinoma of the thyroid in a young woman with gross tracheal and laryngeal deviation clearly seen.

responsible for calcitonin secretion and levels of this hormone may be elevated. A strong familial tendency exists (APUD-OMA). All members of a family should be screened. Spread to regional nodes is common. There may be multiple endocrine neoplasia (MEN) with this, e.g. phaeochromocytoma. Medullary carcinomas are usually slow growing and may be associated with MEN IIa and IIb: phaeochromocytomas and/or parathyroid tumours.

Treatment. The entire gland should be excised because the C cells are scattered and the disease is often multicentric. Radiotherapy and chemotherapy may also be used.

Anaplastic carcinoma

This usually occurs in patients over the age of 60 and is often fatal within 5 years. The tumour typically undergoes rapid enlargement with symptoms of stridor or dyspnoea and often causes regional lymph node enlargement although these can be obscured by massive primary disease (Fig. 310).

Treatment is mainly palliative although radical surgery and RT occasionally produce a good result.

Key Point

The recurrent laryngeal and superior laryngeal nerves are at especial risk during thyroidectomy.

CHAPTER TWENTY NINE

Emergencies

....

STRIDOR

Stridor means noisy breathing. It may be inspiratory or expiratory or biphasic. Inspiratory stridor is usually due to an obstruction at or above the vocal cords. Expiratory stridor is usually from the lower respiratory tract (e.g. wheeze). Two-way stridor is usually due to severe obstruction or disease of large airways such as the trachea or main bronchi.

STRIDOR IN CHILDREN

Newborn infants or children in the first year of life presenting with stridor should always be assessed clinically, with a full history and examination if the infant's condition will permit. The commonest cause of stridor in this age group is *laryngomalacia* (also called congenital laryngeal stridor) but this is not the only cause of this symptom.

Assessment of an infant with stridor

N.B. If a child is cyanosed, it may be impossible to go through the following routine.

History. The phase of stridor and its relationship to activities such as crying, swallowing, etc. are important. Does the child only become distressed and blue when active or when in certain positions? (suggests laryngomalacia). Is the cry weak or otherwise abnormal? (suggests vocal cord palsy). Is the problem exacerbated by feeding? (suggests vascular ring or tracheo-oesophageal-fistula in the neonate). Is the problem only apparent in conjunction with an upper respiratory infection and is the stridor biphasic? (suggests congenital subglottic stenosis).

Examination. Look at the child at rest before moving or handling it. Once a baby starts to cry, it may be impossible to study its resting pattern for some time! Then ask the mother to move the baby into different positions, such as face-down and supine, and again study its respiratory pattern and level of distress. A transcutaneous oximeter is

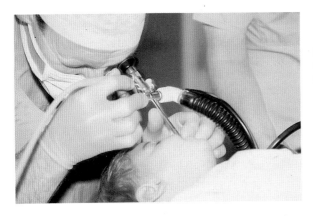

Fig. 311 An infant undergoing rigid bronchoscopy to determine the cause of stridor. The surgeon is using an endoscope with a Hopkins rod light source.

helpful here if available. Then examine the whole child looking for any evidence of external congenital abnormalities before examining the throat. Look inside the mouth and check the palate, tongue, and lower jaw. Always try to watch the child being fed and auscultate the trachea and chest.

Investigations. Oximetry is invaluable. A plain lateral neck X-ray and a P–A chest film should always be obtained if the child's condition permits. This may be the best pointer to confirm a foreign body. A contrast swallow will identify any vascular abnormalities compressing the trachea (e.g. double aortic arch).

Examination under anaesthetic (Fig. 311) is an essential tool for any child whose diagnosis remains in doubt, but this requires the utmost skill and close co-operation between surgeon and anaesthetist. *Any infant being subjected to rigid laryngoscopy and bronchoscopy for stridor may need an urgent tracheostomy to establish or maintain an airway.*

ACUTE INFECTIVE STRIDOR

In the young child this may be due to croup (laryngotracheobronchitis) or acute epiglottitis. These are potentially life-threatening and may need to be managed intensively in hospital. Occasionally the symptoms may be difficult to separate from an inhaled foreign body (see Table 29).

Table 29. Differential diagnosis of acute paediatric stridor

	Group	Epiglottitis	FB
Age	under 2 yrs	over 2 yrs	any age
History	slow	rapid	rapid
Temperature	elevated	elevated	normal
Cough	typical bark	none	spasms

Croup

This is usually of slower onset, with evidence of general systemic upset. It occurs mostly in young children (under 2 years). The child will usually have inspiratory stridor and hoarseness with a typical barking cough. It is viral in origin and cases may therefore occur in clusters.

Acute epiglottitis

Usually of much quicker onset and tends to occur in slightly older children, i.e. 2 years and over. Rapid progression of stridor with drooling of saliva often occurs. This condition is due to infection with *Haemophilus influenzae*. It is occasionally seen in adults.

Both of these disorders may require intensive management with humidification, continuous oximetry to monitor oxygen saturation, and sometimes emergency intubation or tracheostomy. Ampicillin or chloramphenicol are widely used. *Restlessness and tachycardia may be signs of hypoxia.*

A lateral neck X-ray may be helpful in demonstrating a swollen epiglottis but a very sick child may require immediate management. Tracheostomy may be required if intubation fails but this is also fraught with danger.

Foreign bodies

Foreign bodies may be inhaled and may be very difficult to identify clinically in small children. Bronchoscopy may be required to exclude this as a cause of mediastinal shift or localized pulmonary collapse.

> **Key Points**
>
> 1. Avoid distressing a child with stridor as this causes increased oxygen consumption.
>
> 2. If in doubt about a child with stridor, it is best to refer the child to hospital as soon as possible.
>
> 3. Children have small airways and high oxygen demands so the effects of stridor may be dramatic.
>
> 4. Children with epiglottitis may develop airway obstruction rapidly. Insertion of a spatula into the mouth may precipitate acute airway obstruction in a child with epiglottitis and is best avoided.

STRIDOR IN ADULTS

Stridor in adults rarely progresses as rapidly as in children. Although they may suffer from infection, the main worry is laryngeal cancer. The diagnosis can usually be made by indirect or fibreoptic laryngoscopy but expert assessment is required.

TRACHEOSTOMY

This is a procedure to *relieve airway obstruction or to protect the airway* by fashioning a direct entrance to the trachea through the skin of the neck (Table 30). The tracheostomy may be temporary or permanent and is

Table 30. Indications for tracheostomy

Indications	Example
Relief of upper airway obstruction—Actual	Inhaled foreign body Large laryngeal tumour Acute infection in a child
Relief of upper airway obstruction—Potential	After major mouth or pharyngeal surgery
Protection of lower airway—Actual	Overspill of saliva in head-injured patients
Protection of lower airway—Potential	Patients requiring artificial ventilation due to respiratory paralysis, e.g. drugs, Guillaim–Barre syndrome

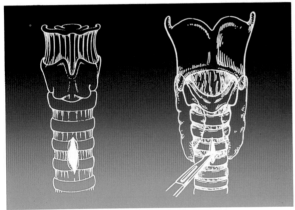

Fig. 312 A diagram illustrating the site of a tracheal incision for a tracheostomy. Note that the cricoid cartilage and first tracheal ring are inviolate.

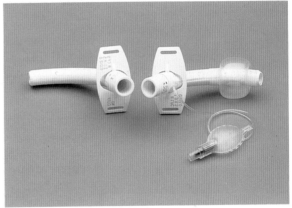

Fig. 313 Silastic tracheostomy tubes. The one on the left is without a cuff. The tube on the right shows a cuff inflated and is suitable for patients undergoing artificial ventilation—e.g. in intensive care.

nowadays usually done following endotracheal intubation. Emergency tracheostomy—when the patient is *in extremis* and the larynx cannot be intubated—is a very difficult procedure and should not be embarked upon lightly. Alternatives exist such as inserting a large intravenous cannula into the cricothyroid membrane which lies in the midline just below the Adam's apple (*cricothyroidotomy* or *mini-tracheostomy*).

The standard operation involves a transverse neck incision, separation of the strap muscles and displacement or transfixion of the thyroid isthmus which overlies the trachea. The trachea is incised (Fig. 312) and initially cannulated with a cuffed plastic tube (Fig. 313) which may be changed for an uncuffed plastic or silver tube 2–3 days later.

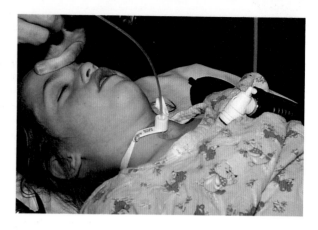

Fig. 314 An infant undergoing suction of secretions which were causing airway problems.

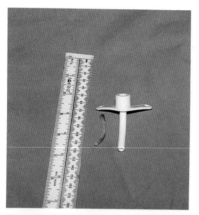

Fig. 315 The tube from the patient in Fig. 314; it was changed immediately when suction failed to improve the airway and this large cast of dried secretions was found in the lumen!

Care of tracheostomies

Highly skilled nursing is required to ensure the lumen of the tracheostomy tube does not become obstructed. Regular humidification and suction (Fig. 314) are essential. If a patient becomes distressed, always check that pulmonary secretions are not crusted in the lumen by inserting a suction catheter. If this fails then change the tube (Fig. 315). Patients who require artificial ventilation or protection of the airway from overspill of saliva need a cuffed tube and cannot change to an uncuffed plastic or silver tube. The latter have the advantage that it may be possible for the patient to speak normally if the tube has the appropriate modifications (see Table 31).

> **Key Point**
>
> Incorrect placement of a tracheostomy tube is a real risk for the inexperienced (Fig. 316).

Fig. 316 Two common ways of misplacing a standard tracheal tube. The misplacement into the pre-tracheal tissues shown on the left is particularly dangerous and more likely to happen if inserted by the inexperienced.

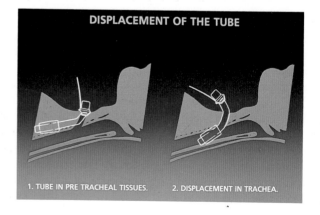

DISPLACEMENT OF THE TUBE

1. TUBE IN PRE TRACHEAL TISSUES.   2. DISPLACEMENT IN TRACHEA.

Table 31. Pros and cons of different types of tracheostomy tubes

Uncuffed tubes	Cuffed tubes
Allow secretions to be inhaled	Prevent aspiration of secretions
Do not touch tracheal wall	Fits against tracheal wall
Suitable for long-term use	Not suitable for long-term use
Air can pass through the vocal cords	Air cannot pass through the vocal cords
Not suitable for ventilation	Suitable for ventilation

CRICOTHYROIDOTOMY

Crycothyroidotomy or mini-tracheostomy is becoming increasingly popular. Its main value is as an emergency access point to the airway (as above). It is also advocated for airway access to facilitate aspiration of bronchial sections in patients with poor cough—especially those in intensive care. This procedure was popular in the early twentieth century but was discouraged by Chevalier Jackson in the 1920s due to the high incidence of laryngeal stenosis. In current practice it is probably wise to limit cricothyroid access to selected cases for short periods.

> **Key Point**
> Cricothyroidotomy is not a substitute for tracheostomy.

QUINSY

This is an abscess in the peritonsillar region. It is usually due to streptococcal infection and causes severe pain with trismus. Examination shows a grossly asymmetric soft palate with displacement of the uvula away from the side of the quinsy.

Treatment is by aspiration or drainage coupled with high doses of penicillin. Some surgeons practice emergency tonsillectomy for this but the procedure is technically difficult. Following a quinsy, some surgeons recommend interval tonsillectomy (about six weeks later). In fact the risk of repeated attacks is not great unless the patient has a history of recurrent tonsillitis or of previous quinsy.

NECK SPACE INFECTIONS

Ludwig's angina

This is an anaerobic infection of the floor of the mouth and sub mandibular triangle. The infection encompasses both sides of the mylohyoid muscle causing distension below the mandible and also elevation of the floor of the mouth. This causes the tongue to be displaced upwards with consequent dysphagia and airway blockage.

Treatment is by drainage and intravenous antibiotics.

Fig. 317 A parapharyngeal abscess presenting in the right upper neck in a young man.

Parapharyngeal abscess

This is usually due to tonsillar infection or occasionally lower jaw dental disease. Swelling is seen in the upper part of the neck behind and below the angle of the mandible (Fig. 317). Examination of the mouth reveals medial displacement of the tonsil on the affected side only. Treatment is by drainage and intravenous antibiotics.

Retropharyngeal infection

Retropharyngeal infections are uncommon and usually occur in children especially under 2 years of age. Treatment is by drainage. This condition is rare in adults and is usually related to tuberculosis of the cervical vertebrae.

> **Key Point**
>
> Only the retropharyngeal abscess is drained through the mouth. All other abscesses are drained externally.

LARYNGEAL TRAUMA

This is uncommon but should always be considered in accident departments or intensive care units when dealing with patients with multiple injuries. Look for surgical emphysema of the neck and abnormal crepitus of the laryngeal framework. The voice will be abnormal but patients in this condition are often unconscious. Early intubation and surgical reconstitution are recommended.

FOREIGN BODIES

Both children and adults may be affected. Children will swallow anything but coins are common. Adults usually suffer from food impaction, e.g. bones and badly chewed meat (Figs 318–320). This is especially common in edentulous adults who may even swallow their dentures—which are usually not radio-opaque.

Clinical presentation

The history of clinical symptoms is paramount in deciding the presence or absence of an ingested foreign body. When the patient has clear recall of the ingestion of the item the diagnosis is easy. However, this is not always the case! Pain, dysphagia and drooling of saliva may follow foreign body ingestion.

> **Key Points**
>
> **1.** Fish bones almost always lodge in the tonsils/base of the tongue and are usually not visible with plain X-rays.
>
> **2.** A normal plain X-ray does not exclude a foreign body.
>
> **3.** Always believe the patient (or parent) who gives a history of foreign body ingestion.

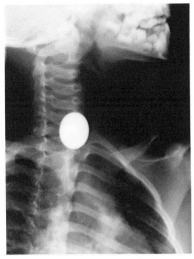

Fig. 318 A coin in the upper oesophagus of a young child. This is one of the commonest foreign bodies seen in children.

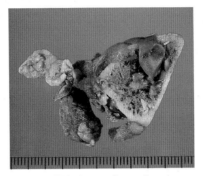

Fig. 319 This huge piece of pork chop was impacted in the oesophagus of a mentally retarded teenager.

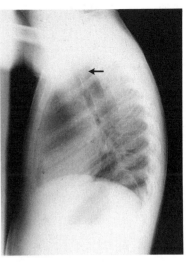

Fig. 320 A lateral chest X-ray showing an intra-tracheal pin (arrowed) which has been inhaled by a child.

Fig. 321 Some of the instruments used for rigid endoscopy under general anaesthesia including an oesophagoscope above and a pharyngoscope below with biopsy forceps in between.

Examination

Salivary pooling and pain are revealed on palpation of the neck. Radiology may be helpful occasionally but is not critical. Specialized studies may be performed in cases of doubt—either a barium swallow, tomography of the neck, or CT scanning.

Treatment

Management in all suspected cases is removal of the foreign body, usually by rigid endoscopy (Fig. 321).

Key Point

The symptoms of foreign body ingestion are more important than the results of specialized investigations.

Index